D0640202

Routledge Introductions to Environment Series

Environment and Business

Alasdair Blair
and
David Hitchcock

London and New York

First published 2001
by Routledge
11 New Fetter Lane, London EC4P 4EE

Simultaneously published in the USA and Canada
by Routledge
29 West 35th Street, New York, NY 10001

Routledge is an imprint of the Taylor & Francis Group

Typeset in Times and Franklin Gothic by Keystroke, Jacaranda Lodge, Wolverhampton
Printed and bound in Great Britain by Clays Ltd, St Ives plc

British Library Cataloguing in Publication Data
A catalogue record for this book is available from the British Library

Library of Congress Cataloging in Publication Data
Blair, A. M. (Alasdair McMillan), 1950–
Environment and business / Alasdair Blair, David Hitchcock.
p. cm. — (Routledge introductions to environment series)
Includes bibliographical references and index.
1. Industrial management—Environmental aspects. 2. Environmental policy.
I. Hitchcock, David. II. Title. III. Series.
HD30.255.B53 2000
658.4′08—dc21 00–038256

ISBN 0–415–20830–0 (hbk)
ISBN 0–415–20831–9 (pbk)

To Gordon Lloyd Blair (1918–1997) and George Henry Hitchcock (1924–1999). Two men who never met, who never even knew each other but who through the support and guidance to their sons made this book possible.

Contents

Figures

Tables

Boxes

Series editor's preface *Environment and Society titles*

The modern environmentalist movement grew hugely in the last third of the twentieth century. It reflected popular and academic concerns about the local and global degradation of the physical environment which was increasingly being documented by scientists (and which is the subject of the companion series to this, *Environmental Science*). However it soon became clear that reversing such degradation was not merely a technical and managerial matter: merely knowing about environmental problems did not of itself guarantee that governments, businesses or individuals would do anything about them. It is now acknowledged that a critical understanding of socio-economic, political and cultural processes and structures is central in understanding environmental problems and establishing environmentally sustainable development. Hence the maturing of environmentalism has been marked by prolific scholarship in the social sciences and humanities, exploring the complexity of society-environment relationships.

Such scholarship has been reflected in a proliferation of associated courses at undergraduate level. Many are taught within the 'modular' or equivalent organizational frameworks which have been widely adopted in higher education. These frameworks offer the advantages of flexible undergraduate programmes, but they also mean that knowledge may become segmented, and student learning pathways may arrange knowledge segments in a variety of sequences – often reflecting the individual requirements and backgrounds of each student rather than more traditional discipline-bound ways of arranging learning.

The volumes in this *Environment and Society* series of textbooks mirror this higher educational context, increasingly encountered in the early twenty-first century. They provide short, topic-centred texts on social science and humanities subjects relevant to contemporary society-environment relations. Their content and approach reflect the fact that each will be read by students from various disciplinary backgrounds, taking in not only social sciences and humanities but others such as physical and natural sciences. Such a readership is not always familiar with the disciplinary background to a topic, neither are readers necessarily going on to further develop their interest in the topic. Additionally, they cannot all automatically be thought of as having reached a similar stage in their studies – they may be first-, second- or third-year students.

The authors and editors of this series are mainly established teachers in higher education. Finding that more traditional integrated environmental studies and specialized texts do not always meet their own students' requirements, they have often had to write course materials more appropriate to the needs of the flexible undergraduate programme. Many of the volumes in this series represent in modified form the fruits of such labours, which all students can now share.

Much of the integrity and distinctiveness of the *Environment and Society* titles derive from their characteristic approach. To achieve the right mix of flexibility, breadth and depth, each volume is designed to create maximum accessibility to readers from a variety of backgrounds and attainment. Each leads into its topic by giving some necessary basic grounding, and leaves it usually by pointing towards areas for further potential development and study. There is introduction to the real-world context of the text's main topic, and to the basic concepts and questions in social sciences/humanities which are most relevant. At the core of the text is some exploration of the main issues. Although limitations are imposed here by the need to retain a book length and format affordable to students, some care is taken to indicate how the themes and issues presented may become more complicated, and to refer to the cognate issues and concepts that would need to be explored to gain deeper understanding. Annotated reading lists, case studies, overview diagrams, summary charts and self-check questions and exercises are among the pedagogic devices which we try to encourage our authors to use, to maximize the 'student friendliness' of these books.

Hence we hope that these concise volumes provide sufficient depth to maintain the interest of students with relevant backgrounds. At the same time, we try to ensure that they sketch out basic concepts and map their territory in a stimulating and approachable way for students to whom the whole area is new. Hopefully, the list of *Environment and Society* titles will provide modular and other students with an unparalleled range of perspectives on society-environment problems: one which should also be useful to students at both postgraduate and pre-higher education levels.

David Pepper

May 2000

Series International Advisory Board

List of abbreviations

ACORN	a classification of residential neighbourhoods
ADAS	Agricultural Development and Advisory Service
BATNEEC	best available technology not entailing excessive cost
BOD	biochemical oxygen demand
BPEO	best practicable economic option
BPM	best practicable means
BSE	bovine spongiform encephalopathy
BTCV	British Trust for Conservation Volunteers
CADW	Welsh Historic Monuments
CANPO	commercial arm of a non-profit organization
CAP	Common Agricultural Policy
CART	Conservation Amenity Recreation Trust
CBA	cost benefit analysis
CBI	Confederation of British Industry
CER	corporate environmental report
CFC	chlorofluorocarbon
CJD	Creutzfeldt Jakob disease
COSHH	Control of Substances Hazardous to Health
CPRE	Council for the Protection of Rural England
DETR	Department of the Environment, Transport and the Regions (1997)
DG XI	Directorate General 11 of the EU Commission
DOE	Department of the Environment (1970–97)
DTI	Department of Trade and Industry
EA	Environment Agency
EC	European Community
EDLRB	environmental damage limitation and repair businesses
EIA	environmental impact assessment
EIS	environmental impact statement
ERT	European Round Table of Industrialists

ESA	environmentally sensitive area
EU	European Union
FHB	Farm Holiday Bureau
FoE	Friends of the Earth
GATT	General Agreement on Tariffs and Trade
GCC	Global Climate Coalition
GDP	gross domestic product
GIS	geographical information system
GMO	genetically modified organism
GNP	gross national product
HMIP	Her Majesty's Inspectorate of Pollution
HSE	Health and Safety Inspectorate
HSEP	health, safety and environment policy
IoD	Institute of Directors
IPC	integrated pollution control
ISO	International Standards Organization
LCA	life cycle analysis
LETS	local exchange trading systems
LFA	less favoured area
MAFF	Ministry of Agriculture, Fisheries and Food
MNC	multinational corporation
NCC	Nature Conservancy Council
NEPA	National Environmental Policy Act
NFU	National Farmers Union
NGO	non-government organization
NIC	newly industrializing country
NIMBY	Not in my back yard
NNR	National Nature Reserve
NPO	non-profit (or not-for-profit) organization
NRA	National Rivers Authority
OFWAT	Office of Water Services
PEST	political, economic, social, technological
PPP	polluter pays principle
RSPB	Royal Society for the Protection of Birds
RSPCA	Royal Society for the Prevention of Cruelty to Animals
SERA	Socialist Environment Research Association
SME	small and medium-sized enterprises
SSSI	site of special scientific interest
TGI	target group index
TUC	Trades Union Congress
UDC	underdeveloped country

UNEP	United Nations Environment Programme
VOC	volatile organic compounds
WTO	World Trade Organization
WWF	Worldwide Fund for Nature (formerly World Wildlife Fund)

Introduction: the environment and business

This book has come about with several overall objectives in mind. We have attempted to explore the range of ways in which the various relationships between business and the environment operate. We are aware that these relationships have changed through time and so have attempted to describe something of how the changes have come about. Through developing our ideas and researching this book we have become aware of the complex interactions between business and the environment. We suspect these are more numerous than may at first appear to many individuals and more numerous than is suggested by much of the current literature. We have looked at the nature of these relationships across the range of primary, secondary and tertiary industries. Our examples have been drawn from all three of these sectors to show the range of different business relationships with the environment. Finally, we have examined how the current (and past) literature has discussed these relationships. We hope we have been able to add something by way of a different perspective to the overall understanding of what has clearly become an important area of study.

The text has been written with certain readers in mind. We started from the premise that the reader has some basic knowledge of the environment or will be consulting other texts in the series. We have taken the view that the environment consists not only of the system of physical and tangible elements evident in the landscape such as forests or heaths but also consists of the less obvious system of physical and biochemical processes which interact with and influence the tangible elements. We have assumed the reader may not have such a systematic understanding of business. We spend a little time examining how businesses operate as seen through the eye of the environmentalist. We briefly describe the relevance of such elements as labour inputs, capital acquisition and funding, revenue and cash flow, marketing, information gathering/ processing and business law in the environmental context. While these topics will be all too familiar to the student of business, previously such students will probably not have seen these activities viewed through 'green lenses'. For environmentalists, geographers and planners these topics are likely to be entirely new. What we see as important are the various relationships between business and the environment. We have used diagrams extensively to complement the text description in an attempt to ensure the reader's understanding of the various relationships. Within each chapter we have included what we have termed a 'Discussion point'. These are questions and exercises designed to get the reader thinking about the issues raised in the book. Some can be attempted merely from reading the appropriate chapter, but others require further independent work, for example, accessing a website or carrying out a survey. Many exercises are open-ended and are intended to encourage the reader to relate what is in this book to the real world.

Chapter 1 introduces the basic concepts and thinking behind the book. We begin by examining the relationship between the environment, business and environmentalism, leaving signposts where we will examine ideas in greater detail later. We explain what we mean by the three terms (environment, business and environmentalism) and then examine the various relationships between them. The points are made that the environment can either (a) react to business activity or (b) can deterministically influence business.

Finally, in Chapter 1 we look at how those with alternative social and political views believe business operates with regard to the environment. This section questions the virtues and alternatives to a market economy centred on economic growth and examines the paradox between the will of environmentalists to be in harmony with the environment and the business community's mission to exploit it for financial gain.

Chapter 2 traces the development of business, industries, the physical environment, environmentalism and social attitudes to the environment over the last two hundred years. The chapter catalogues developments showing the way in which the five factors have developed in relation to each other. The chapter concentrates particularly on the way in which business culture and environmentalism have evolved together through time, particularly in recent decades.

Chapter 3 examines the way in which an individual business looks upon the environment in the light of present-day business operations and the prevailing business ethos. The chapter examines the behaviour of individual firms and how they consider the environment in business terms. Some businesses regard the environment as an asset either in terms of its physical resources or as a marketing opportunity. Other businesses view the environment as a cost. There are costs to be incurred through:

- compliance – having to comply with prevailing legislative measures or social demands;
- loss of business which results from environment pressures not having been taken fully into account;
- choosing voluntarily to incur costs which are not required by legislation. Businesses will do this to either preserve an image, enhance their market share or to forestall stricter legislation in the future.

Businesses believe that incurring costs in the short term for whatever reason will bring corporate benefits in the longer term. Such businesses will adjust their operations and structures to accommodate these costs. They may take on additional labour to carry out such operations, they will adjust their financial structures and possibly their marketing operations. A third group of businesses look on the environment as an externality. They export all their costs which means the costs must be borne by others, very often the general public.

Chapter 4 examines the way in which external pressures with an environmental dimension are applied to business in general. These are the pressures applied to a business over which they have no, or limited influence, including such things as legislation or changing climate. The chapter examines the way in which those pressures related to the environment influence the thinking of businesses. We have adopted the often used approach of examining these outside influences in terms of a PEST analysis, examining the way in which Political, Economic, Social and Technological influences come to bear. In addition to the usual group of four factors we also include the physical environment as a fifth factor exerting what many will see as a deterministic influence on a business. We are aware that any pressures cannot be seen as falling exclusively into one of these five categories and that there are ripple effects from one to another. For example, a social pressure on the company to behave in a certain way might well have political or technological implications. How the reader sees these forces will depend to some extent on their background. The economist may be conditioned to see everything in terms of economic pressures, the lawyer or sociologist in terms of social pressures, the engineer in terms of technological pressures and undoubtedly the environmental scientist in terms of physical pressures.

The chapter will include consideration of such techniques as environmental impact assessment and environmental audit as tools to assess the consequences of these pressures. Chapter 4 is about how business reacts to the costs, pressures and externalities imposed upon it.

In Chapter 5 we move on from the pressures to look at the pro-active ways in which businesses can respond to environmental influences. We are aware of the old management adage that what one business regards as a threat another regards as an opportunity. This chapter is about how businesses see the environment as an opportunity. It examines the possibility of a win-win game where a business can fulfil its commercial objectives, while at the same time enhancing the environment. Although the chapter is about how businesses use the environment as an asset, the contention that environmental damage might ensue from even the most 'green' of businesses is also considered.

Chapters 6, 7 and 8 are about how the issues, forces and complexities described so far work themselves out in reality in different industrial sectors. The different industrial sectors make different demands on the environment and have different impacts. The primary industries, examined in Chapter 6, use resources directly and are dependent on the supply of such resources. The secondary industries (Chapter 7) transform the resources and their processes are likely to have significant environmental impacts as a result of manufacturing processes. Some texts on business and the environment are almost exclusively concerned with the manufacturing industries of the secondary sector and so much of the ground covered here may be familiar to many of you. The tertiary sector (Chapter 8) uses the products of the other sectors. They create environmental impacts in more subtle ways which have become more important as this sector has grown to dominate economic activity in advanced economies. In each of these three chapters we look at the sector in general before going on to examine one industry where the complexities involved can be developed in depth. For each sector we have explained how compliance, reaction and pro-action work and what forces and attitudes come into play.

Chapter 9 is about those businesses that have resulted directly from environmental concern. These can be seen as the most pro-active of all the businesses. They begin as environmental concerns and issues that have been turned into businesses.

Chapter 10 gives the opportunity for speculation and considers how the interrelationships between business and environment are likely to develop in the future, using the format adopted by other texts in the series.

We have attempted to reflect a variety of different political and social standpoints. We have championed neither a technocentric nor an ecocentric view of the world. Nor have we taken a Marxist, classical economic or any other economic stance. We neither accept business and environmentalism are cosy bedfellows nor that they are implacably opposed to each other, regarding either standpoint as an over-simplification of the true position.

However, none of us come to a subject without our philosophical baggage or education (depending on how you view such things) and what we already know of a subject. We implicitly accept the conventional orthodoxy of a modern commercial business setting. We assume that economies usually evolve, starting from a predominantly primary-based economy, moving through a manufacturing stage and finally ending up as a service sector economy. We acknowledge the views of some left-wing thinkers such as Bookchin and Foreman (1991), but regard such an approach as a interesting model but one which is not in general accord with how we see the world operating.

We see a pluralistic society in which there are stakeholders and opinion formers who together have a considerable influence on the policies which emerge in our society. Of course, the use of such terminology in itself determines from where we are coming. We see business as a very varied series of commercial activities. The businesses we will be discussing range from large multinational corporations to individuals. They are drawn from the primary, secondary and tertiary sectors of industry. Just about the only characteristics they have in common is they seek to fulfil a business objective within a commercial context.

In our approach we hope that we have been distinctive in how we have gone about our task. We have taken the view that to understand the complete relationship between business and the environment it is necessary to understand both the way in which the environment influences business as well as the ways in which the various industries change the environment. Many texts have set themselves the task of explaining either the role of environmentalism in influencing business or the technocentric explanation of the impact of industry on the environment – but not both. We have also attempted to examine the complete range of business attitudes to the environment and to understand businesses' standpoint without making any value judgement about their attitude. We have been able to identify that businesses may look on the environment as either an asset or as a cost or as an externality – concepts we discuss in depth in Chapter 3. For those companies who regard the environment as an asset, we have analysed the pro-active steps they take to use the environment to fulfil their business objectives.

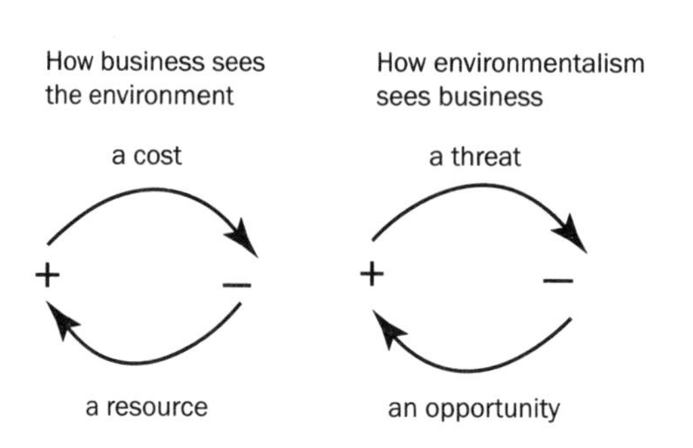

Figure 0.1 ***Different ways of seeing the environment***

There is a spectrum of attitudes between seeing the relationships in a positive or a negative light (Figure 0.1). This may be likened to how a pint glass with half a pint of liquid in it is viewed. The optimist sees this as a half-full glass: the pessimist a half-empty glass. In the same way, business may see the environment as either a cost or as a resource. Similarly, the environmentalist may see business activities as either a threat or as an opportunity.

We have not restricted ourselves to what many have regarded as those sectors of the economy which have had the most dramatic impact on the environment, namely, chemical and petrochemical companies, mining concerns and 'green' manufacturers. We have attempted to examine the complete spectrum of business. By examining the primary, secondary and tertiary industries separately we hope we have been able to identify a range of different interactions not usually considered. We have gone out of our way to examine business activities that are often overlooked as industries and where the environmental impacts are often not even considered. These include retailing and construction through to the environmental influences of the post-industrial economic activities that are now beginning to impact on the nature of the countryside. We have paid particular attention to agriculture since we believe this is an important business activity and one that has a close inter-relationship with the environment.

Finally, we have looked at the way in which business and the environmental movement in some respects are moving closer together. While other texts have examined the way in which business has taken on the thinking and methods of the environmentalists, particularly in relation to such techniques as environmental audits and impact assessment, we think it distinctive that this text looks at the ways in which non-profit-making organizations such as the National Trust or RSPB have taken on business methods to raise money to allow them to fulfil their environmental objectives. This novel relationship between business and environment is such that it has led us to identify a new species of business organization – the CANPO (the Commercial Arm of a Non-Profit-Organization).

1 Environment and business: the nature of the relationships

- **The basic elements – business, environment and environmentalism**
- **Impact of business on the environment**
- **The environment influencing business**
- **Business cultures and ethics**
- **The globalization of business**
- **Environmental and political standpoints**

> Simple answers rarely suffice for complex problems; environmental and natural resources problems are not an exception.
>
> (Tom Tietenberg, 2000)

> The environment is in the process of becoming a major new competitive area for business. Understanding the impact of the relevant issues and responding to the resulting opportunities and threats will become an important source of competitive advantage.
>
> (Elkington and Burke, 1987)

One of the main aims of this first chapter is to familiarize the reader with the relationships which are to be explored in this book. The relationships in themselves are relatively straightforward.

First, business activities have an impact on the environment. This occurs through pollution, the modification of ecosystems, introducing exotic species and chemical compounds, genetic engineering and irreversible physical changes – the sorts of impacts widely reported in the popular press and television. The ways in which land has been used, particularly for agriculture and forestry, are a reflection of the way in which the commercial pressures of business have changed the physical landscape. Business affects the environment in a variety of ways both deliberate and accidental. Most of the effects are usually seen as detrimental, although increasingly we see some business activity improving the environment. For example, a number of museums and country houses have been opened to the public where the restoration of former days of glory have halted decades of decline and reinstated previous, albeit not entirely natural, surroundings. However, many firms who claim to be improving the environment are in reality only mitigating against the most injurious impacts of what is occurring. Hence the practice of planting trees and landscaping around industrial plants is not improving the environment but merely reducing the detrimental effects of building the plant in the first place.

Second, but less obvious, are the impacts the environment has on business. Changes in climate have led to changes in business activity. Depleting resources or physical limitations on the use of some materials have changed patterns of exploitation and their uses. Changing environmental conditions have led to the birth of new industries. Although less widespread and less covered by the literature, these changes are still worthy of our attention.

Associated with this first pair of relationships is a second pair, which concern the relationship between business and environmental thinking and perception. Third, business activities are influenced by environmental concerns. It is hard to identify a business where environmental concerns have no impact on its operations. For most businesses, environmental issues are less important than other external influences such as the state of the economy or the behaviour of its competitors. For some, which are close to the environment such as agriculture or tourism, the impacts can be at the top of their agenda. Environmental legislation, customer attitudes to the environment along with direct (and more subtle) actions by environmental groups are among the most obvious influences which can be grouped together under the umbrella term of environmentalism. Environmentalism has a considerable impact on many individual businesses. Businesses are increasingly aware of public pressure to have due regard for the environment and now many have corporate policies that are either sympathetic, or pro-active, to environmental aims. The way in which business behaviour has been influenced and modified, as a result of the influence of environmentalism, is something that will be developed more than other relationships in this book.

Fourth, business has worked hard to influence public opinion about the environment. While many such actions can be seen merely as window dressing, we shall see that many business opportunities have been seized to pro-actively develop products and services resulting in environmental improvement. Business has actively promoted some of its activities in terms of the environmental benefits they have brought. The public debate over the introduction of genetically modified (GM) foods has not been one-sided. Business, and in this case one company in particular, Monsanto, has done much to publicize the environmental benefits from the use of such products. Business has actively lobbied the media and government to win over the hearts and minds of the population over a number of issues. Claims and counter-claims by business and their detractors have done much to muddy the environmental waters. It has brought about an even greater need, however, to understand the true nature of the relationship between business, the environment and environmentalism.

The basic elements – business, environment and environmentalism

This book, in essence, is about the reciprocal relationship between business, the environment and environmentalism and this is where we begin our analysis. A business exists to fulfil its mission statement and to further its commercial objects, which are often expressed in financial terms and rarely make any reference to the physical environment. However, a business cannot ignore the environment, which it may see as either a positive or a negative factor. It may see the environment as a marketing opportunity, or as an opportunity to diversify into new areas of activity. In either case the environment will be seen as an asset which will enable the business to achieve greater commercial success.

By contrast, another business may see the environment not as an opportunity, but as a cost factor. How this influence operates is examined in Chapter 3. The environment exerts a cost on the commercial operations of the firm. The firm sees this as a legitimate part of its operations and adjusts its structures accordingly. A parallel can be drawn between the way in which environmental and labour costs are incorporated into an organization's structure. Both are legitimate costs for which the organization makes allowance.

Some businesses regard the environment (or at least certain aspects of it) as an externality – a cost which is passed on to others. They make no allowance in their structures to accommodate the impacts their business makes on the environment. Any environmental costs are exported to others. The difference between costs and externalities is subtle but important.

This book is essentially about the interaction between the environment and business. In doing so it draws a subtle but fundamental distinction between *business* and *industry*. Although commonly the words 'industry' and 'business' are used interchangeably, we believe the differences are significant enough to merit separation in our discussion.

An industry is seen as the collection of firms who operate essentially the same series of processes that result in a related set of products (whether tangible products or services) that a third party wishes to buy. Products and services originate in, and are delivered to, particular locations. Their environmental impact may be felt at the point of production, the point of consumption, the transport link between the two or any combination of the three. By convention, industries are divided into primary, secondary and tertiary industries. Primary industries include fishing, forestry, agriculture and the extractive industries (essentially, the quarrying and mining of minerals). They all involve the collection, harvesting and exploitation of resources directly produced by physical processes. Secondary industries are the manufacturing industries. They take raw materials and by a variety of processes produce tangible goods by adding value to the raw materials. The tertiary industries produce services, for either individuals or for other organizations. Some authors have extended this tripartite division to the quaternary and quinary sectors, by subdividing the tertiary industries. Such distinctions are not felt necessary here since such sectors are not distinctive in their relationship with the physical environment. However, the way in which primary, secondary and tertiary industries inter-relate with the environment are seen as being sufficiently different to warrant separate analysis.

Business is the range of commercial organizations and their activities that characterize the way in which trading is conducted in a capitalist economy. It is concerned with the management of factors leading to the production of goods and services and the ways they are brought together and delivered to a customer. Business refers to the organizations that carry out such operations, how they are structured and what drives and motivates them. It is about the financing, marketing, the legal framework, collection and transfer of information, the human skills, training and behaviour of the individuals who carry out these trading patterns. These activities cannot be tied to any one industrial sector or to any one geographical location and so the ways they interact with the environment are harder to identify. Nevertheless, we feel it important to distinguish between business methods that can be common to all types of industry and specific industrial sectors, which have distinct sets of relationships with the environment and environmentalism. Both perspectives are necessary to obtain a rounded view of the overall relationship between business and the environment. Before the development of the capitalist economy and the concept of the 'market', business, as we understand it today, was very different. Industry before the Agricultural Revolution and the development of the factory system was radically different in both its scale and operation and in its relationship to the environment. We are therefore taking as our starting point for modern industry the quantum leap forward about two hundred and fifty years ago which came with the Industrial Revolution, when the technology and methods of manufacture as we know them today emerged.

Any starting point must be regarded as arbitrary but the above has served to make a clear distinction between industry, which is about people using resources to produce products or services, and business, which is about people using and making money. The distinction can be illustrated by looking at a business such as the oil multinational Shell. The business consists of two parent companies, Shell Transport and Trading Company plc (the British part) and The Royal Dutch Petroleum Company (the Dutch part). Together they retain the shares in group holding companies which in turn hold the shares in a myriad of service and operating companies which include subsidiaries in many countries. These companies operate in all three industrial sectors. Extraction of crude oil is part of the primary sector, refining is in the secondary manufacturing sector and distribution and retailing are in the tertiary sector.

Environmentalism is about the preservation and enhancement of the environment. In order to achieve their aims an environmental organization may either:

1 put pressure on the business community to mitigate the effects of environmental damage. This can be done by:
   (a) influencing public opinion directly; or
   (b) influencing government either directly through lobbying or indirectly through influencing public representation to government; or
   (c) influencing businesses directly. Such environmental organizations include high profile pressure groups such as Greenpeace and Friends of the Earth, as well as a host of lesser known organizations. The way in which environmentalism affects business is considered in Chapter 4.
2 adopt business principles to further their environmental aims. Hence

an environmental organization, such as the RSPB, has incorporated marketing, financial management, retailing and promotional activities into its operations to raise the necessary finance, publicity and a favourable image to enable it to fulfil its environmental mission. How such organizations have adopted business methods is explored in Chapter 9.

Environmentalism in itself is a complex movement. Broadly, environmentalism can be described as the movement concerned with the protection of the environment whose proponents – environmentalists – consider it should have a primary influence on society's development. The protection of the environment can be considered as coming within the custodianship of three distinct groups:

1 governments or official agencies with statutory obligations to protect the environment;
2 environmental pressure groups who broadly believe they can add something to that being done by statutory means;
3 individuals who have an opinion on the environment and who may take positive steps to protect it. Just as each of these groups have a range of views about society so they have contrasting attitudes to the relationship between society and the environment. Hence, environmentalists represent a range of political, social and ethical standpoints. As individuals we can all be pigeon-holed into one of the following viewpoints where we either:
   (a) believe the environment will be saved by technological developments and that business will be the saviour of the environment;
   (b) believe we should be living in harmony with our environment and that the only way forward is through sustainable development;
   (c) reject the idea that the market can solve any problems and look to alternatives to the capitalist market economy in the future;
   (d) maintain that the environment is irrelevant.

The final word we need to examine at this stage is *environment*, which to most readers will broadly mean the 'surroundings'. However, it is a word that is interpreted in different ways. We have used the word environment throughout as it is used by geographers, planners, earth and biological scientists to include both the tangible elements and the processes of the physical world. These are the tangible physical resources of the globe, the atmosphere, the land surface, rocks and minerals, plants, animals and the seas. It is also taken to include the physical and biological processes that go on close to the surface of the earth independently of any input from humans, although they are constantly modified by the actions of man. These processes result in the movement of water, energy, biological organisms and other essential compounds within the physical world. These ideas we know will be familiar to many readers, but this definition is in contrast to what business students tend to understand by the word 'environment' – as used in many business and economics texts. For them the word is the same in that it means the external elements of the surroundings. However, the business paradigm understands the surroundings not in terms of a physical entity but rather as such factors as the behaviour of the market, the international economy, the legal framework, competitor activity and consumer behaviour. Where there is any possible confusion between the terms, we have used the term 'physical environment' for what elsewhere we have simply called 'the environment'. In general the context of the discussion will make it clear what we mean.

At this stage it is perhaps worth pointing out that both business – by which we mean the business community as well as the collection of individual companies – and environmentalism are both pro-active in that they bring about change by their actions. Both are constantly changing in how they function and even their overall objectives do not remain the same for ever. We will see in Chapter 2 that both have changed considerably over the last two hundred years. The environment, by contrast, takes no conscious decisions itself. Instead, it facilitates the self-maintaining natural physical processes or homeostasis. These tend towards physical equilibrium under 'natural' circumstances or are driven by the natural variations that have occurred through geological time.

Impact of business on the environment

From the outset it should be clear that some types of business are more closely linked to the environment than others. For example, agriculture, heavy chemicals and holiday companies promoting tourism to natural unspoiled areas of the world are closely aware of the environment – albeit in very different ways. Other companies, such as hairdressers, computer software developers or assemblers of television sets, have little to do with the environment directly and one assumes consideration of the impacts on the environment do not come up as a matter of regular discussion at their management meetings. This book concentrates on those sectors that have a close relationship with the environment. It also focuses on those organizations which have either taken a particularly pro-active response to the environment, or by their woeful inaction have had a noticeable detrimental impact on the environment. All these responses, however, are conditioned by the ownership and structure of the business, by how the business believes society views the environment and the business's attitude to the environment.

We are going to develop a model in which businesses are both influenced by the environment, on the one hand, while at the same time having some impact on it. Few businesses are totally free from the influences of the environment, although many have no significant impact themselves. These influences come through pressures to comply with environmental legislation or, through market pressures, to appear to be green. For some businesses, however, there is a much higher dependence on the environment. We shall see this dependence extends to all three sectors of industry: primary, secondary and tertiary. For the present the example of agriculture will suffice, since farming is clearly dependent on environmental factors – most obviously climate, hydrology and soil fertility. The relationship between dependence and impact is shown in Figure 1.1. The transhumance pastoralist is dependent on the environmental conditions but makes relatively little impact. By contrast, factory farmers have largely insulated themselves against the vagaries of nature but produce a much greater impact on the environment due to waste products, pollution, visual intrusion, etc. An organic farm depends on the environment to a greater degree since it cannot use artificial fertilizer or pesticides and it must make use of natural inputs to achieve its aims. An organic farm still has impacts on the environment but they are more benign than a factory farm. A farm museum has little direct dependence on the environment and little direct impact, although indirectly it could contribute to pollution and congestion caused by its visitors.

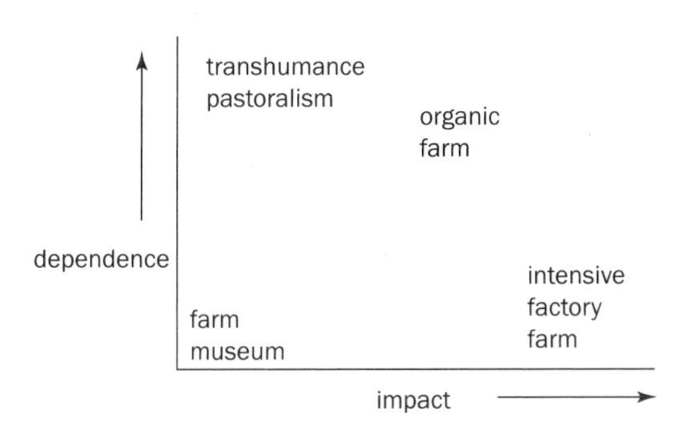

Figure 1.1 ***Impact and dependence on the environment***

Superimposed on the inherent nature of the type of business relationship with the environment is the behaviour of the individual company. The individual business is not a passive player in this relationship. Most can do little to influence the external environment – it is the province of only the largest of organizations that they can sway the market or influence government legislation. The individual company will impact on the environment in a variety of ways. In later chapters we will examine different strategies firms can adopt which will either increase or minimize any impact. We would expect that as the impact on the environment increased, so a firm would become more concerned with its relationship with the environment, although this may not be the case.

We make the point that business has two effects on the environment and environmentalism. First, it changes the environment either directly or indirectly as a result of its commercial activities. Second, it influences environmental legislation and the attitudes of government, society at large and other organizations. Businesses are not only subject to influence from other organizations but also from the collective pressure exerted by individuals in society.

The environment influencing business

In Chapter 4 we consider more fully the ways in which the physical environment influences business operations. These are mainly concerned with either systematic changes in environmental conditions, e.g. global warming, or as a result of sudden and unexpected 'natural' events such as floods, storms or earthquakes. What is important to realize is that business behaviour is adjusted to expected variations in physical conditions. Just as companies successfully operate within an envelope of changing economic conditions – economic growth, inflation, unemployment, etc., so they operate within an envelope of varying physical conditions. Calamitous events present major problems and disruptions to their operations – be they physical or economic calamities. Because the rate of systematic physical change is slow (way beyond the planning cycle of most companies), many environmental changes have little impact on their activities. Hence if they occupy a coastal location and sea level is rising at a rate measured in centimetres per century this is not going to be of great concern to a company focusing on its business strategy over the next two years. As physical conditions change slowly, so the company will adjust accordingly.

Only in exceptional cases do systematic environmental changes influence business. Farming in the Mid-West of the USA during the 1920s and

1930s was influenced by changes in physical conditions – namely, the creation of the Dust Bowl. The herring industry of the North Sea has had to bow to the impact of the shoals of herring no longer being so numerous. Such examples are, however, comparatively rare, since most environmental change is comparatively slow. It is only in the area of research and development where business activity looks sufficiently far into the future (thirty or forty years) that environmental change has time to take effect. Hence major oil companies are actively pursuing alternative sources of energy for the time when oil reserves finally become in very short supply. Shell are carrying out extensive work in the field of solar energy and BP in the field of wind power. Biotechnology companies, such as Monsanto and Zeneca, are anticipating the time when new crop varieties will be required to meet new environmental conditions. Such developments, however, are the exception and to the casual observer it appears that while business is changing the environment, the environment is not changing business practices.

Business cultures and ethics

It is necessary to understand that different firms have different cultures, which influence their attitude to the environment. How they go about achieving their mission is influenced by executives in the firm and by stakeholders, where the beliefs and personalities of individuals can take on a pivotal role. Sometimes, in small firms, the relationships are simple with few players. In large organizations there is usually a greater degree of complexity. Individuals may have become stakeholders due to their views on the environment. Non-executive directors may become board members due their environmental viewpoints or expertise. Charitable institutions may make similar appointments as trustees. Banks have in recent years become more aware of environmental matters. Some banks have declared they are only willing to loan to companies whose environmental and / or ethical viewpoints are within particular parameters. The Co-operative Bank has perhaps taken the highest profile stand on such issues. Institutional investors have recently come under scrutiny over particular investments. Some investors have become acutely embarrassed when it has become apparent where they have placed their investments. Non-executive directors are other stakeholders who may well bring with them strong environmental views and who may have an important role in shaping the approach taken by the company. There are also examples where environmental protesters have acquired shares in a company. They have then used the public forum of the shareholders' meeting to bring about an orchestrated publicity event to embarrass the board, who are duty bound to attend such meeting. As a result of such action companies may change their policies.

However, away from the public gaze, executives within companies develop particular strategies and images for the business. Although beyond the scope of this text, it is worth noting that the environmental stance and attitude of companies are rarely mere accidents, but rather a set of conscious decisions taken and driven forward by executives from within an organization.

Business displays a range of ethical standpoints that may, from the viewpoint of the environmentalist or the general public, seem to be at best lukewarm, and at worst hostile, to the needs of the environment. Those firms who possess a traditionally good or bad record of environmental awareness have most to gain from considering their position. Those who have no record of environmental responsibility in the past can make rapid progress. Those who have always had such considerations on their agenda can capitalize on the structures and mechanisms they already have in place. Those in the middle have a more difficult task since they have done the easy things, but to go much further they must radically change their thinking and procedures. Business activity has a very widespread impact on the environment because of its ubiquitous nature. There is nowhere on the planet which is not in some way affected by business activities, directly or indirectly. Even Antarctica, the last great wilderness, has its waters trawled by commercial fishing fleets and its coasts visited by tourists. Is it inevitable that business and environment should come into conflict? Or is it possible to reconcile the two?

No modern business would admit to being anything other than responsible and ethical in its activities and environmental care takes its place alongside respect for human rights, workers' rights and community responsibility as a principle that ought to be central to a company's concerns. As a chairman of Shell has said: 'We believe that without principles, no company deserves profit. Without profits, no company can sustain principles' (Moody Stewart, quoted in Brown, 1998). Here is a crucial point, profit without principles is immoral, but the company cannot afford to have principles at the expense of profit or it will soon cease to be viable. There is inevitably a trade-off between principle and profit. Neither purely altruistic nor purely unprincipled businesses are likely to survive – the first because they will go bankrupt, and the second because no one will do business with them.

How do businesses view the environment in practice? One indication is the type of environmental policy that they adopt. Even if they do not have a formal policy, it is possible from their actions (and inaction) to see how they treat the environment. Increasingly, the more environmentally aware business has a formal policy and produces a Corporate Environmental Report. Ketola (1997) maintains that we can examine company environmental policies on the basis of values, visions, goals, means and the general strategic management approach which will also apply to environmental matters. A dynamic progressive firm, which actively seeks novel change, is likely to be more receptive to environmental improvements than a moribund company which seeks to suppress change and discourage initiative. With the environment creeping up the agenda, those firms prepared to incorporate an environmental policy pro-actively are likely to succeed more than those who would like to pretend there is not a problem and that if they ignore it it will go away. Table 1.1 shows the range of possible responses by businesses to the environment and their associated strategic management.

Table 1.1 shows how different aspects of policy inter-relate to produce an overall approach of a firm to the environment. The least green firms

Table 1.1 ***Companies' responses to environment: the range of policies***

Environment concern	*Environmental values*	*Energy visions*	*Environmental goals*	*Environmental means*	*General strategic management*
1 no concern	very strongly technocentric	oil age	stability	defensive	suppress change
2 below average	strongly technocentric	oil and gas	reactive	defensive and cooperative	adapt to change
3 average	modified technocentric	traditional mix	anticipatory	cooperative	seek familiar change
4 above average	ecocentric	progressive mix	entreprenurial	cooperative, threatening	seek related change
5 high	biocentric	renewable	creative	threatening	seek novel change

Source: Adapted from Ketola (1997)

adopt the posture of the ostrich and stick their heads in the sand in the hope that the danger will go away if they pretend they cannot see it. Thirty years ago this was probably the most common attitude of business. Gradually, with varying degrees of concern, businesses have conceded that the environment is an issue to be addressed by them and have moved in different ways to accommodate it in to their strategies. Some make minimal change and are essentially reacting to perceived threats, whilst an increasing number of firms try to anticipate change but strictly on their terms. In Ketola's words, they seek familiar change. The environment is acceptable if it can be incorporated into the conventional cost structure. This is essentially the position of the eco-modernists, who try to incorporate ecological thinking into the business framework. More radical approaches, as shown by those with ecocentric and biocentric values, are more pro-active and seek to use the environment as an opportunity and to deliberately seek change in their business orientation. These are the firms who deserve the label green businesses. As we shall see later in the book, these businesses are increasing in number and show the way to a more sustainable economy.

The globalization of business

We now turn to the international dimension of the relationship between business and the environment. It is widely recognized that most large businesses now operate at the international level. Globalization is seen as an increasing feature of twentieth-century business. Multinational corporations (MNCs) transcend national boundaries in their operations. Economic changes that generally go under the heading of globalization are at the heart of the contemporary transformation of business. Globalization is essentially the growing integration and interdependence of the world's economic systems. Globalization of production, services and consumption is a process which has been spearheaded by ever more powerful MNCs who see it as an imperative of growth. Contemporary capitalism has established itself as the dominant economic force, seeing off state socialism as its only serious rival by the late 1980s. Globalized capitalism has required a number of key factors to emerge in order to develop in the way that it has. Organizationally, the multinational corporation is the key player with diversified global production and sales. An MNC owes no particular allegiance to any one country even though they normally originate in the USA, Japan, or Western Europe and because of this they regard investment or disinvestment in any location as a purely business matter. This footloose behaviour of MNCs has been facilitated partly by changes in the attitudes of governments to the rules of international trade and finance and partly by the changing technology of communication exemplified by the Internet which has allowed effective real-time control of far-flung subsidiaries. The result has been what Dicken (1998) has termed a 'global shift'. The old industrialized countries of Europe and North America have experienced de-industrialization, a loss of their old staple manufacturing industries that have been relocated to cheaper sites in Third World countries. They have been replaced by service industries and increasingly the knowledge-based industries and 'e-commerce' based on the Internet. Newly industrializing countries (NICs) such as South Korea and Taiwan which had become the locations for much manufacturing capacity in the 1970s are now in their turn being challenged by cheaper countries in their periphery.

Seeking ever greater productivity gains, firms have attempted to increase economies of scale by merging and forming alliances across countries and continents. For example, in the first two months of 2000 alone, AOL merged with Time Warner to form the world's largest media company, Vodaphone took over Mannesmann to become the world's largest mobile phone company and Smith-Kline-Beecham merged with Glaxo-Wellcome to form a pharmaceutical giant. Such mergers frequently lead to job losses, since there is no need, for example, for two corporate headquarters buildings. Despite this danger of increasing unemployment governments have been willing accomplices in the development of contemporary capitalism. The post-Second World War approach of governments towards the economy involving tight control of trade, planning and state-owned industries have been swept away by the growth of economic neo-liberalism, associated most strongly with the governments of Margaret Thatcher and Ronald Reagan in the 1980s. The most significant developments which have assisted globalization have been the liberalization of trade through the successive rounds of the General Agreement on Tariffs and Trade (GATT) negotiations which culminated in the Uruguay Round of 1994 which set up the World Trade Organization as a new regulatory body to oversee the furtherance of free trade, and the deregulation of the global financial system which allowed the unfettered movement of finance across the globe. Privatization of publicly owned industries in energy, water, transport and telecommunications has also been widespread and many of the newly privatized companies have sought international markets. As part of the globalization process there has been the deliberate fostering of a global consumer culture, with common fashions in clothing, fast food, music, TV and sport supplied by globally recognized brands, e.g. Coca-Cola, McDonald's, Nike, CNN, Manchester United. Supermarkets in Britain now directly control production of fruit and vegetables in Africa, Central America and the Caribbean to provide out-of-season produce for consumption in British cities. The result in economic terms has been a tripling of world trade by value, vast growth in foreign direct investments world-wide and huge fortunes accruing to the CEOs of MNCs.

In environmental terms, these changes have contributed to greater despoliation of nature, for example, continued deforestation of South East Asia to feed insatiable demands for wood in Japan, forest clearance in Central and South America to provide cheap ranch land for cattle destined for hamburger chains. Growth of air travel to connect the centres of business have resulted in more pollution as have the increasing Western consumer affluence which demands not just one car per household, but as many as can be crammed into the parking spaces. Growing concern has been expressed that increasing consumer lifestyles world-wide, fostered by the imperatives of MNCs, have contributed to serious environmental problems. These include global warming, acid rain, oil spills and the hole in the ozone layer, as well as a huge variety of local environmental controversies such as road, dam and airport construction. All have helped to boost environmentalism, not just in the prosperous West but also in the developing countries. Globalization of communications has led to more awareness of environmental issues and a growing sense of communality of problems between different regions. There has also been a realization that environmental issues are inextricably intertwined with economic, social and political problems. For example, Watts (1998) shows how the environmental devastation of Ogoniland by the oil industry in Nigeria was intimately connected with the political oppression of the Ogonis, their economic impoverishment and political powerlessness. Issues of human rights, social justice, economic self-determination all found expression in an environmental disaster. The well-known green aphorism 'think globally, act locally' has started to become more appropriate as local groups have started to relate their own concerns with wider global issues. In 1999, a group of Third World farmers travelled round Europe protesting about corporate attempts to control agriculture by patenting seeds which coincided with local protests in Britain over GM foods.

The end of the twentieth century saw much discussion about post-industrial society, postmodernism, the 'end' of everything from science to history and from geography to work. Much of this may have been opportunist capitalization on apprehension about the coming millennium. However, some ideas deserve a little more consideration as they have relevance to any discussion of business–environment relationships. The idea of a post-industrial society gained much credence in the West where industry was visibly disappearing and more and more people found employment in services in the dying days of the twentieth century. There was a frenzy of anticipation about the new 'e-economy', the web-based electronic economy. This was fuelled partly by huge growths in share prices of Internet Service Providers and partly by the media and politicians who seemed determined to encourage the population become 'wired'. In theory, an Internet-based economy could be more environmentally friendly in that it would drastically reduce the need for physical transport of people for work or shopping. However, the goods would still need to be transported to people's homes and delivery vehicles would increase in number. The main environmental benefit of a post-industrial society would seem to be the lessening of energy, waste and pollution from industrial manufacturing. However, this is an illusion, since the factories have simply moved overseas to pollute other areas of the world and their contribution to global warming is just the same. Transferring pollution may take it out of sight but it is still there causing damage on a global scale. Beck (1992) has put forward the theory of the 'risk society' as the next stage forward from industrial society. In industrial society, the rich lived some distance from the mines and factories and did not suffer the costs of pollution and industrial accidents. The costs all fell on the working classes who lived next to the mine or the factory, so that, for example, if there was a mine disaster, the people who suffered were the families who lived in the colliery village, not the shareholders who lived three hundred miles away. In the risk society, however, the scale of activities and possible hazards affects everyone wherever they live, the radioactive cloud from Chernobyl was no respecter of property or national boundaries and who became contaminated was decided by the random action of the unpredictable weather conditions. Beck regards existing institutions as failures in coping with this type of risk and foresees the evolution of networks of new institutions transcending traditional boundaries to regulate and ameliorate risk. His ideas have been criticized as being abstract and the idea that the risk society has replaced the industrial society is over-stated according to Dryzek (1997).

Sustainable development is usually defined as meeting the needs of the present without compromising the ability of future generations to meet theirs. The concept of sustainability can be simply explained as ensuring perpetual renewability of resources, ecosystems and human communities. Extraction of a resource must be balanced by an equivalent replacement, so, for example, if 100 hectares of forest is cut down, an equivalent area of saplings must be planted to replace it, or if fossil fuel must be burned emitting a certain amount of carbon, then the equivalent quantity of carbon ought to be fixed by planting the requisite number of trees. However, while environmentalists are concerned to modify business in order to make the environment sustainable, business is more concerned about the sustainability of its own activities. Sustainability has an ambiguity about it that allows it to be readily adopted by both environmentalists and by business and politicians. They are interested in sustainability in order to be able to continue economic growth. It is easy to see that the two ideas are not quite the same. Business has on the whole embraced the concept of ecological modernization (see below) as the principle ethos for achieving sustainability. Environmentalists have regarded this as a step in the right direction if not necessarily the most desirable way to achieve it. Environmentalists have been often split between what the German Green party describes as Fundis and Realos (fundamentalists and realists). For the green fundamentalist, capitalist business can never be right since its profit-making consumerist principles will inevitably despoil the Earth. But for the environmental realist while business has to be accommodated, the trick is to try to persuade it to do the right thing slowly even if this means compromise. Ecological modernization may not be the best theoretical system from a purely ecological or moral viewpoint but it is the most practical in the present circumstances. Compared to attitudes in the 1960s, business has become more accommodating to environmentalism and environmentalists have begun to appreciate the need to work more closely with business rather than simply to oppose it.

We will see in Chapter 2 how, in Western countries, the relationships with which we are concerned in this text have evolved over the last two centuries. In many countries the developments, which took two hundred years in Britain have been telescoped into a few decades. Looking across the globe today we see countries at all stages of transition. Some still have an essentially primary-based economy, producing and exporting agricultural products and other raw materials, usually minerals. Other countries such as China have become recently 'industrialized', whereby the manufacturing sector has become the most important sector of the economy. Some countries have already developed a strong tertiary economy and now have an industrial structure very similar to that of Western countries, e.g. Hong Kong, Taiwan, Singapore. Table 1.2 shows the industrial structure of selected countries.

Table 1.2 ***Distribution of Gross Domestic Product between industrial sectors in selected countries, 1995***

Region	*Percentage of GDP attributable to*		
	agriculture	*industry*	*services*
World	5	33	63
UK	2	32	66
France	2	27	71
Germany	1	38	61
Japan	2	38	60
Singapore	0	36	64
Korea (South)	7	43	50
Malaysia	13	43	44
Mexico	8	26	67
China	21	48	31
India	29	29	41

Source: World Resources Institute (1998) from data Table 6.1, pp. 236–7

Two features need to be taken into consideration in this oversimplified model of development. First, some countries have moved directly into the service economy. This has occurred where either there are strong international trading links present (as in Dubai) or where tourism (as in some Caribbean or Indian Ocean islands) has developed instead of manufacturing. Second, we see patterns of what might be termed reverse investment taking place. For example, South Korean firms have invested in Northeast England in areas still scarred by the remnants of the earlier manufacturing cycle. Global corporations invest anywhere in the world where suitable conditions can be found.

The reaction to environmental concerns may differ in various parts of the world, even within the same company. A series of factors come together to steer MNCs in different directions. These can be briefly summarized as:

- Companies may have individual parts of their operations with contrasting attendant environmental risks in separate countries. A company may have all its manufacturing capability in a Third World country or, may mine in one country, process in a second and assemble and market finished products in a third.
- Where company operations are further away from their home base, adherence to standards may fall short of stated policies. This results either because managerial control is less or it may be a deliberate policy of a company tacitly supporting the irregular (and cheaper) practices of its subsidiary.
- Environmental standards and monitoring of standards vary in different parts of the world. Hence practices which are legal (and cheaper) in one country may be illegal in another. This may influence a company to move its operations across national boundaries.
- On the positive side, companies are aware of legislation and accepted practice through their spheres of operation and so may well adopt as company policy the best practice throughout their plants and so raise accepted standards world-wide.
- MNCs have the advantage of being able to shift their operations to countries where conditions best enable them to fulfil their mission and have the infrastructure to develop brands with a world-wide image. However, the world has not only become smaller for such companies, it has also become smaller for the media and the pressure groups who are aware of developments. This means that adverse exposure in one country can have a world-wide impact on operations and so increasingly companies are adopting world-wide standards and awareness of the environmental consequences of their actions.

It is dangerous to draw any conclusions about the overall environmental behaviour of MNCs. We shall see in this book that there are a variety of different corporate responses to the environment, of which MNCs exhibit the full spectrum of behaviour. Superimposed on this variety is the fact that different conditions exist for different industrial sectors in different countries.

Environmental and political standpoints

Environmentalism comes in a huge variety of philosophical viewpoints, ideologies, or world views. Dryzek (1997) identifies nine principal environmental 'discourses' (most of which have various sub-types) which represent the main ways people think about the environment and influence their actions towards it. The discourses are:

- survivalism (limits to growth);
- Promethianism (technical fix);
- administrative rationalism (regulatory approach);
- democratic pragmatism (let the people decide);
- economic rationalism (let the markets decide);
- sustainable development;
- ecological modernization;
- green romanticism (eco-radicalism, eco-socialism);
- green rationalism (eco-realism).

Individuals and groups which enter into dialogue with business may hold any (or some combination) of these viewpoints. Although business may not subscribe to most of these discourses (indeed, may regard them as a threat), it needs at least to be aware of them. In negotiating with a community over location of a new factory, it would be helpful to know if they were talking to adherents of green rationalism or green romanticism, for example, to be able to establish some common ground. While it may be of interest to business to be aware of the range of environmental attitudes, relatively few of these discourses have practical application for mainstream business applications.. Table 1.3 shows some principal environmental ideologies and their relationships to business matters. It can be seen that eco-radicalism and eco-socialism advocate deep change to society in order to create an economy in harmony with nature. Eco-radicalism aims to by-pass capitalism altogether and create self-sustaining communities through grass-roots action. It is in a sense opting out of the present system and attempting to operate an alternative type of ecologically sound business. It does exist in small and scattered 'communes', but its capacity to become widespread is unlikely in the extreme under present social and economic conditions. We do not intend to comment further on the various shades of green to be found among deep ecologists, eco-feminists, animal liberationists, Gaians, eco-anarchists or lifestyle greens. Such groups tend to have an anti-business ethic and are more likely to confront rather than cooperate with business and it hardly needs pointing out that business is unlikely in the extreme to adopt their ideology as its own.

Eco-socialists (or eco-marxists) believe that environmental problems are a symptom of the workings of the capitalist system and that structural reform of the system along socialist principles is the answer to the

Table 1.3 ***Business/environmental philosophies***

Type	*Description*	*Comment*
eco-radicalism	alternative thought, deep green ecology, rights of environment, people and species paramount	highly principled, limited to a few scattered communities, unlikely to appeal to the majority as it needs radical changes to lifestyles
	communitarian and utopian, by-pass capitalism and opt out of system	
eco-socialism (eco-marxism)	application of socialist ideas/ principles to environment; change modes of production to benefit people and nature; confront capitalism; state intervention	high principles marred by poor record of socialist states in practice; remaining 'socialist' states increasingly 'capitalist' in attitudes to business
eco-liberalism	incremental, slow reform of the system to create gradual improvement, legislation against worst environmental excesses, but seeks changes in attitude through education, tax and voluntary means	most successful of the types so far; ideas adopted by most states; slow change means environment still deteriorating, despite genuine improvements
eco-modernism	incorporation of environment into the business agenda; environment on business terms; helps business first, environment second	increasingly acceptable to the establishment of business and government; fits 'Third Way' ideas
eco-efficiency	the practical arm of eco-modernism uses technology to solve problems, increases efficiency, cuts costs, make profits and solves problems	appears to solve problems, but does not deal with fundamentals, technical fix which may simply delay rather than solve crises

Source: Terms after Welford (1997), descriptions and comments: authors

ecological crisis. They seek fundamental change to society through a top-down approach via state intervention. The distinction between eco-socialism and eco-marxism is not altogether clear, but like their parents, socialism and Marxism, their respective adherents seem to regard themselves as distinct. Both feature the application of socialism to the environment with changes to the means of production, altering of needs and redistribution of wealth. Unlike the classic socialist regimes of the past they seek to benefit the environment, not to ignore it. Despite Western socialists' disavowal of the old Soviet-style of socialism, they suffer from a credibility problem in that, since the general demise of socialist regimes in Eastern Europe, it has become apparent that socialist governments, in practice, had been even more devastating to the environment than their capitalist counterparts (Lewis, 1992). Those avowedly socialist governments that survive, like China, seem to have done so by adopting capitalist business methods. Western countries that had adopted some socialist policies in the post-war period, such as nationalization, generally abandoned them in the 1980s. A socialist environmental agenda is advocated for example by the Socialist Environment Research Association (SERA) which acts as an environmental think tank for the British Labour Party (SERA, 2000). Marxist analysis as a means of explaining the development of the world economy still has many academic adherents and has begun to be applied to environmental issues (Braun and Castree, 1998). Business, however, operates within the dominant neo-liberal capitalist system and it is hard to imagine Shell, Sainsbury's or Microsoft run on either eco-radical or eco-marxist principles.

Eco-liberalism has been the most widespread approach with its emphasis on gradual reform of the existing system and incremental change through the tax system, codes of conduct, regulation and education. In fact, this is the mainstream philosophy practised by governments and business in most of the western world. Ecological modernization is a more recent offshoot of eco-liberalism that is an attempt by business to incorporate the environment into its agenda on its own terms. The environment is regarded as a legitimate and unavoidable area of concern, but the concern is primarily for the well-being of the business not the environment. Ecological matters are incorporated into business in so far as they are compatible with business requirements. Welford (1997) has termed this the corporate 'hijacking of environmentalism'. Unable to ignore or refute the concerns of the public, its shareholders or the government over the need to 'do something', business has attempted to bring the environment on board, but in ways which do not fundamentally threaten business operations or pursuit of profit. It is regarded as a necessary compromise by many environmentalists and as anathema by the eco-radicals. The success of ecological modernization is shown by the way in which primarily environmental organizations, such as voluntary groups, have begun to adopt business methods and business language to run their affairs. As we shall show later in the book, the development of business to support environmental objectives has been a striking feature of such organizations in recent years and promises to be even more significant in the future.

Ecological modernization involves a restructuring of the economy along more environmentally sound principles, but it does not require any radical change to the current economic or political system. It is therefore an unthreatening ideology which can be accommodated by business without changing their essential goals and structures. It is appealing to business because it offers the chance to make not just environmental improvements and thereby gain the applause of governments and public, but to make cost savings and improve balance sheets. Ecological modernization is seen as good business practice, summed up in the phrase 'pollution prevention pays'. Unlike the regulatory approach which deals with 'end of pipe' problems, ecological modernization seeks to resolve pollution and waste by designing it out of industrial processes. It is, for example, cheaper and more effective to prevent contamination of groundwater than to clear up the mess later. In addition, ecological modernization has also stimulated a pollution prevention and control industry, thus furthering commercial opportunities.

Ecological modernization concentrates on the practical and achievable and so has much more appeal to business people than many other environmental ideologies, which, while they have grand visions, have little practical advice on how to achieve them, or can point to few successes by their adherents. Ecological modernization is characterized by the themes of partnership between industry, environmentalists and government to solve problems. It is therefore a problem-solving paradigm which regards economic growth and environmental protection as compatible. This is a positive, inclusive approach that does not seek to demonize business and is obviously more appealing to business than many of the other environmental ideologies that portray it as the villain. It has attracted some environmentalists since it does offer real scope for progress. However, ecological modernization is not to everybody's taste. It treats the environment as a commodity to be fitted into the business worldview and it has little to say about how to treat non-economic aspects. It is also has an incremental, case by case approach and assumes that the aggregation of individual companies' actions will add up to an overall environmental benefit. Green radicals are uneasy with ecological modernization because it has deflected attention away from their fundamental criticisms of the capitalist system towards immediate problem solving. Although it may result in the more efficient use of resources, it still stimulates the growth of consumption and may therefore be regarded as postponing the ecological crisis rather than resolving it. It has also weakened the radical environmental movement by attracting away those environmentalists who have despaired of any progress through the other ideologies. Ecological modernization has a practical plausibility which most other environmental ideologies lack. It stresses the use of technological innovation, capital and management techniques to increase efficiency while achieving environmental objectives, for example, reducing energy use or recycling waste into useful products. It is claimed as the answer to resolving the contradiction between business growth and environmental needs A good recent example is the study by Von Weizacker, Lovins and Lovins (1998), whose title alone: *Factor Four: Doubling Wealth, Halving Resource Use*, describes the idea concisely. There is an almost evangelical flavour to the writings of the proponents of ecological modernization (which is matched by the eco-radicals and eco-socialists), but unlike their deep green opponents, their message is received with favour by the business community, since it is not threatening to capitalism, but actually seems to offer a viable, even profitable way to resolve the environmental problems of business.

Summary points

- This first chapter has introduced the basic relationships of the book. These are in two pairs, first, how business has changed the environment, and how the physical environment has influenced business activity. Then, how environmentalism has influenced the way business is conducted, and how business has reacted to the environmental lobby.
- It has clarified some basic terms: what we mean by the physical environment, the difference between the environment and environmentalism, and the difference between business and industry. We have examined the reciprocal influences of environmentalism and business. It has looked at the main ideological and philosophical standpoints taken by business in its approach to

environmental matters and by the environment movement in its attitudes to business.

- Having read Chapter 1 it is hoped that the reader will appreciate the various approaches which might be taken to these issues and to understand that the relationship between business and the environment can be seen from a range of different viewpoints.

Further reading

One of the major problems of the environmental literature is its breadth and range of political viewpoints. For an investigation of the relationship between business and the environment a good starting point is the series of papers by D. Smith (ed.) *Business and the Environment* (London: Paul Chapman, 1993) which in addition to the political, legal and social dimensions to the issues includes a particularly useful account by Dave Green on the role of accountancy to the green agenda. Another text which provides a comprehensive and well-balanced account of many of the issues which are discussed here is to be found in S. Eden *Environmental Issues in Business* (Chichester: Wiley, 1996). For an understanding of the way in which business has enhanced the environment the texts by D. Welford and R. Starkey (1996) *Business and the Environment* (London: Earthscan, 1996) and J. Elkington and T. Burke *The Green Capitalists: How to Make Money and Protect the Environment* (London: Victor Gollancz, 1987) are valuable contributions. For a very different viewpoint A. Dobson *Green Political Thought* (London: Unwin Hyman, 1990) advances the ideas of ecologism as a left-wing alternative viewpoint to conventional thinking, which is extremely well argued and informative. B. Braun and N. Castree *Remaking Reality: Nature at the Millennium* (London: Routledge, 1998) gives a good idea of the Marxist and postmodern approaches to environmental matters. J.S. Dryzek *The Politics of the Earth: Environmental Discourses* (Oxford: Oxford University Press, 1997) provides a useful overview of the wide varieties of environmental ideologies.

Discussion points

1 What is ecological modernization and why has business adopted it more enthusiastically than any other environmentalist ideology?

2 Evaluate the significance of globalization to the environment.

Exercise

Speculate on how any of the following businesses: Microsoft, Sainsbury, Shell, Disney, might be run on the lines of: (a) eco-radicalism; (b) eco-socialism; (c) ecological modernization. Pay particular attention to commercial viability and to environmental impact.

2 The changing relationships through time

- The changing environment
- Doing business over the last two hundred years
- Developments in the three industrial sectors
- Birth and growth of environmentalism

> [T]he Yellowstone, Yosemite, Sequoia etc. – Nature's sublime wonderlands, the admiration and joy of the world. Nevertheless, like anything else worthwhile, from the beginning, however well guarded; they have always been subject to attack by despoiling gainseekers and mischief makers of every degree from Satan to Senators, eagerly trying to make everything immediately and selfishly commercial with schemes designed in smug-smiling philanthropy, industriously and sham piously crying 'conservation, conservation, panutilization', that man and beast may be fed and the dear Nation made great.
>
> (John Muir, 1912)

> Historically the relationship between economic development and environmental protection has been seen as one of mutual antagonism.
>
> (Gouldson and Murphy, 1997)

This chapter shows the ways in which the environment and business have changed over the last two hundred years. The last two hundred years have seen fundamental changes to the physical environment – probably greater than any other two hundred year period since the dawn of civilization. The landscape has changed during this period as a result in the shift of power from agricultural to industrial interests. Business has experienced even more dramatic change to the point that what now passes for business would be unrecognizable to our predecessors in 1800. Although commercial transactions took place before then, they were on a different basis from today both in terms of how companies were capitalized and how the market operated.

The origins of the Industrial Revolution have their roots in a series of events which took place in Western Europe, and in particular in Britain, just over two hundred years ago. Other areas of the world followed what began in Western Europe. In some newly industrializing countries the time frame in moving from a rural, largely agricultural, to an urban, largely industrial, society has been compressed into a matter of one or two decades. Many industries were established when environmental concerns, awareness, knowledge and attitudes were different from those of today. Other industries' fortunes have waxed during periods of lower environmental concern and now are on the wane, e.g. the coal industries of Western Europe. Other businesses began at a time when environmental concerns were different from those today and so take time to adjust to new environmental attitudes, e.g. supermarket chains. Businesses have not only had to come to terms with new attitudes and values but also with new technologies, new machines, new chemical compounds and new biological strains. Each in turn has had a profound impact on the environment over the last two hundred years.

Society's attitude to environmental matters has evolved with the other changes. These changes should be seen alongside striking changes in the values, ethics, morals and beliefs held by society, which have themselves largely been a consequence of industrialization and urbanization. New lines of communication, the changing role of the media, education, and a shift from the need for survival to a desire for an improved quality of life – at least in Western societies – has changed our views of what is important. The rise of environmentalism has influenced the views of the individual and these have put pressure on business to change. Two hundred years ago there were no environmentalists as we know them today. Gradually, their views have shaped society's attitude to the environment and to the impact that business has had on the environment.

Business, environmentalism and society have evolved in parallel. By tracing their historical roots it helps us to understand how the present relationships have come about and reinforces the lesson that history has a habit of repeating itself. Studying the historical roots of the subject can prove valuable for a number of reasons. First, the forces which ultimately bring about change have often operated for many years before the full impact of the change is recorded, e.g. the shift in political power from agricultural to industrial interests in nineteenth-century England. Second, sometimes some momentous or seminal event initiates change at some later date, e.g. the publication of Rachel Carson's *Silent Spring* in 1962 is seen as a major turning point in environmental thinking. Third, we shall see that history has a habit of repeating itself. For example, the two occasions when environmental awareness was at its greatest – in the 1840s and 1850s and then in the 1960s and 1970s – corresponded to periods of relative prosperity and free thinking. Finally, the implications of any industrial development can never be anticipated from the outset, e.g. nobody could have foreseen the social, industrial and environmental impacts that were to follow when the first motor cars took to the road. These factors explain many of the tensions that exist between business and the environmental movement since the interpretations that can be put on the same event may be different in the two camps.

The changing environment

In 1800 the world was much closer to its natural state than it is today. Despite people having lived on every continent for millennia, much of the world was sparsely inhabited and in huge areas of North and South America, Siberia, Sub-Saharan Africa, Australasia and the Pacific islands a true wilderness still existed. Although there had been former civilizations, such the Aztecs, the Mayans and the Incas, little long-term effect on the environment had resulted. This may be due to the fact that population numbers were not large and that the populations had had an ethos of co-existence with nature. Even in Europe the impact of industry and agriculture was still limited and even the largest cities were small by today's standards. Huge areas of untouched forest, virgin grassland and undrained wetlands abounded. Wildlife was already in retreat (the last wolf having been shot in Scotland in 1746) but in the sparsely settled lands it still teemed, e.g. the buffalo on the North American prairies. The pre-industrial seas and oceans were vast when compared to the small amounts of activity which were restricted to a few miles close to land. Agriculture relied entirely on human and animal power and used only natural recyclable materials for fertilization. No artificial pesticides existed. Agriculture was largely a subsistence, but sustainable, way of life and the transport links did not permit agricultural produce to be moved over large distances. However, this should not be seen as a 'Golden Age' since life was hard, subject to natural forces, disease was rife and life expectancy low. Nature had not yet been tamed.

Over the last two hundred years all this has changed due to agricultural change and population growth associated with urbanization and industrialization. The percentage of the world covered by both arable land

and permanent pasture, which is often enclosed, has increased. The twentieth century has seen moves towards commercial cropping and away from traditional husbandry, including shifting pasture and temporary tillage (Briggs and Courtney, 1985).

In the last two hundred years the world population has risen dramatically from an estimated one billion in 1825. It broke the two billion barrier in 1930, three billion in 1960, four billion in 1977, five billion in 1989 and reached six billion in 1999. The population of England and Wales increased from 10.1 million in 1811 to 26.0 million in 1881 (Checkland, 1982). This increase was far from evenly distributed with 94 per cent of the increase being absorbed by the towns (Lawton, 1978). This century the population has continued to rise, reaching 32.5 million by 1901 and 49.8 million by 1991. More people have meant a greater environmental impact and greater requirements for food, water, building materials and energy as the cities have become progressively more populous world-wide. As people move off the land so those remaining are required to produce even more food associated with a greater ecological footprint, in the terminology of William Rees (Smith *et al.*, 1998).

The third major agent of change has come from industrialization leading to more pollution, increasing exploitation of the world's natural resources and less contact with nature. The interdependence between industrialization, increased population, urbanization and the commercialization of agriculture has brought about a position proclaimed by McKibben in 1990 as the 'end of Nature', in the sense that the natural world had become irrevocably altered by industrial processes. Although the foundations of environmental degradation were during the Industrial Revolution, it has been in the last part of the twentieth century that global environmental destruction came to be seen as a serious problem.

Although the effects of these changes are slow, few if any areas of the natural environment have been untouched by their impacts. The composition and dynamics of the atmosphere have been changed by the emissions of greenhouse and other gases and by particulate emissions. This has led to a change in the energy balance of the atmosphere and a shifting of climatic belts. Emissions of CFCs have thinned the ozone layer of the upper atmosphere resulting in more UV radiation reaching the surface, leading to an increase in skin cancer rates. The effect of heat, pollution and the changes to the natural air circulation in cities has led to regimes of urban climates characterized by increased sudden rainfall and higher temperatures. Excavations, agricultural activities and construction works have changed the very land surface as everything from houses to large dams and reservoirs have been built. Overall, increased pressure on less stable land has resulted in extensive land instability and landslides (Coates, 1981). This has led to erosion resulting in increased sediment in rivers. It is no over-statement to say that the overall shape of the land surface has been changed over the last two hundred years. Flat coastal areas have not been spared from considerable modification. They are extensively used for both habitation and industrial use. This has meant coastal protection and land reclamation works have been used extensively. These efforts to bring stability to coastlines have often had the opposite effect, resulting in further instability along the coast bringing about the need to build further extensive protection works (Walker, 1997).

The hydrological regime of the land has been extensively modified. Lakes have become acidic (Battarbee *et al.*, 1997), the nutrient balance of streams has been changed and watercourses are clogged with weed. This in turn has led to eutrophication and the creation of algal blooms. Agricultural land drainage and extensive sewerage schemes in urban areas coupled with the canalization of rivers have changed the flood regime of many rivers. These changes, when coupled with the constriction of floodplains by urban developments and changes to regimes from dams, reservoirs and a host of water abstraction schemes, result in significant modifications to the surface water hydrology (Hollis, 1997; Goudie, 1993). Groundwater levels have been reduced due to increasing demands from agriculture, industry and public water supplies and pollution from industrial, domestic and agriculture sources means many groundwater supplies can no longer be considered wholesome.

Natural wetland areas have been drained, often to realize their agricultural potential. This has led to the environmental and wilderness value of such sites having been lost, particularly in the last half century. The forests of the world have undergone significant changes. Natural forests, with their considerable species diversity, have been felled, sometimes to be replaced by commercial plantations. Natural grasslands have been subject to an even more dramatic contraction. Many have been changed into enclosed permanent, or semi-permanent pasture, or have been turned over to arable production. The contraction of the extensive grasslands in North America and the plains of Africa have resulted in dramatic consequences for the wildlife of the areas. Animal ecology elsewhere has also been altered. Many species have become extinct. And biological diversity has been reduced in many areas. This has made it easier for population explosions of pests to take place. Exotic species have been introduced into many areas, resulting in a decline of native species and disruption of the natural ecological balance.

The last ecosystem, which has been modified less than most but which still has been subject to change, is that of the ocean. This has not been subject to the same pressure as much of the land surface. However, the size and perceived invulnerability of the oceans have led to a tendency for them to be used as a dustbin for hazardous and unwanted wastes, where the long-term impacts of such actions have not been assessed. As affluence has spread, so the demands upon natural resources have mushroomed. Humanity has become an urban species more and more divorced from daily contact with nature. Paradoxically, as this happened, people began to value nature more and more and have sought to protect and preserve what remains.

Doing business over the last two hundred years: a case study of the UK

At the beginning of the nineteenth century, business was dominated by numerous small, usually family-owned, companies. The raw materials were secured locally as was the labour force and the capital. It was Keynes who has been accredited with noting the nineteenth-century morality whereby it was acceptable for the rich to remain very rich, so long as they reinvested their wealth in their industrial enterprises and retained exclusive control over these companies. In the climate where new markets were emerging, existing markets expanding and where the barriers to entry were low, small firms could flourish and large firms, e.g. Coates' yarns, Wills' tobacco and Cadbury's chocolate, were the exception. Checkland (1982), when examining the sources of initiative that led to the rise of the industrialized society, distinguishes between men of invention and men of business who brought about the dramatic events of the nineteenth century. What is also clear is that the nineteenth century was a time when individuals made a difference. By contrast, the twentieth century has been a time when corporations have made the difference.

By 1870 changes started to become apparent as Britain lost her initial industrial advantage over her rivals, most notably Germany and the USA. Free trade began to be replaced by protectionism as economic growth faltered and limited liability companies became more popular (Pawson, 1978). They provided a mechanism for the financing of very large companies. Second, they provided a way in which the rich could diversify their wealth into a portfolio of different sectors of the economy and, third, they led to the evolution of the professional managerial class. Perkin (1989) observed 'The rise in the scale of organisation is not the cause but the effect and symptom of the rise of a much more complex inter-dependent society.' Perkin pointed out that in 1880 the hundred largest firms in the UK were responsible for less than 10 per cent of national production, which rose to 26 per cent in 1930 and 45 per cent in 1970.

This century two world wars and cycles of depression and growth have shaped the way in which business has developed. The two world wars marked important turning points not only during hostilities, but also in that they triggered a series of changes that meant that business would never be the same again. Some would never recover; for others, particularly those centred on the manufacture of the hardware of war and those of strategic importance at home (including agriculture), it acted as a stimulus. The wars also increased the role of government in industrial planning and compelled competitors to cooperate with each other, both of which made mergers and takeovers after the war more likely.

The Second World War also left an impact on the way in which business was conducted (see Jeremy, 1998). A culture emerged where:

1 Technological developments were being advanced by teams working for well-resourced organizations. This approach which had developed radar (Checkland and Holwell, 1998) and broken German ciphers along with the commitment which was necessary for the production of aircraft, tanks, submarines and munitions during wartime could now be turned to commercial developments.
2 The role of the professional manager (Perkin, 1989) characterized the emergence of new organizational structures and new hierarchies in business. He begs to differ from F. D. Roosevelt's view that the twentieth century is the century of the common man, maintaining that it is the century of 'the uncommon and increasingly professional expert'. While some found it difficult to make the transition from military command to civilian management, the way in which a large number of people could be organized to meet a particular goal influenced the rise of the corporate commercial climate.
3 The importance of logistics, intelligence gathering and the increasingly important role of the management of information led to

the development of new management systems. The techniques and mentality which had been able to put together the strategic planning and coordination required for the D-Day landing were now put to commercial use.

4 The war had also fostered a belief in science and technology as the way in which business was to go. There was a realization that the war had been won by 'boffins' as well as by bravery and weight of numbers and that scientific-based enterprises were to be the way of the future.

In the 1960s and 1970s business increasingly came under the influence of merger waves and of internationalization. Mergers affected all sectors of the economy (including the tertiary sector) and most featured horizontal mergers. Some were takeovers, others had all the characteristics of an inefficient company imploding upon itself, e.g. British Leyland in its various guises. The last forty years has seen increased multinational activity with large British firms taking an interest in foreign concerns and overseas multinationals setting up plants in the UK to gain access to European markets from within the trade barriers of the EU. Electronics concerns (Fujitsu, Sony, Siemens, Phillips) and car manufacturers (Honda, Peugeot, Toyota and Nissan) are examples of foreign-owned manufacturers who have established themselves in the UK. They have tended to be located in the traditional manufacturing areas, based on former coalfields, where government support and subsidies are greatest. Such areas have suffered most from the 'branch office problems' of being susceptible to closures and cutbacks during troughs in the economic cycle. The impact of such ventures in environmental terms can be seen as twofold. On the one hand, such employers have no direct contact with the local environment and so no continuing reason to ensure its well-being. On the other hand, the prosperity that such schemes have brought have enabled inward investment to improve the environment in ways which was not possible without such plants.

Developments in the three industrial sectors

The last two centuries have seen a changing balance in the share of employment, output, productivity and GDP attributable to primary, secondary and tertiary industry. All sectors have seen several revolutions in technology, business structures and methods that have had important impacts on the environment and environmentalism. We need to briefly consider both the legacies of these developments for the present-day relationships between environment and business and historical parallels which may illuminate current issues.

Table 2.1 ***Employment in the three industrial sectors in the UK***

	% of total employed workforce		
Year	*Primary*	*Secondary*	*Tertiary*
1801	43	17	40
1851	25	30	45
1890	16	44	40
1950	5	46	48
1973	3	42	55
1998	2	29	69

Source: Checkland (1982), Perkin (1989) and Jeremy (1998)

The hallmark of a pre-industrial economy is the overwhelming dominance of primary production, particularly agriculture, and the lack of an industrial–urban infrastructure. As countries develop, they undergo a transition from rural-agricultural to industrial–urban and eventually develop a post-industrial-service economy. As can be seen from Table 2.1, even by the nineteenth century the UK had moved a long way along the development trajectory in terms of employment in industry and services.

Primary industries

By 1851 the UK already had half its population living in urban areas and the decline of agriculture was underway. During the eighteenth century the UK underwent the Agricultural Revolution (although some, such as Kerridge (1967), put this at an earlier date) which began the transition from subsistence way of life to commercial business. The application of technology replacing human and animal labour began in this period but did not reach its full potential until the twentieth century with the development of mechanization and chemicalization. It also saw the victory of capitalist business methods with the enclosures (which could be regarded as a form of privatization of communal resources) and the marketing of food to the urban masses. However, the victory of the landowners was short-lived as the agricultural development of overseas farming especially in the Americas and Australasia, soon undercut British farm prices. Increasingly, along with other European countries, British agriculture entered upon a long phase of relative decline, despite very impressive productivity gains and large increases in the volume of output during the second half of the twentieth century. The role of state support, first from national government and then through the CAP of the European Union, was crucial in the prosperity of the farm sector. Initially designed to boost farmers' incomes and increase output, the policy was increasingly seen as an albatross around the Union's neck. It was expensive, failed to stem the decline of small farmers, encouraged over-production by large farmers and was implicated in environmental destruction of landscapes, habitats and wildlife. Agenda 2000 is but the latest of a long series of reforms designed to tame the CAP expenditure, by cutting back price support, channelling help directly to farmers and encouraging diversification and environmental schemes (European Parliament, 1999).

The historical legacy of agriculture lies mainly in the landscapes created over the centuries by farming activities that have become part of the national heritage. These semi-natural landscapes are also wildlife habitats, though years of relentless chemical warfare against pests and weeds have diminished much of their ecological value. It is significant that government subsidy is now being increasingly directed to the preservation and reinstatement of environmental features on farms which only a few decades ago farmers were being paid to remove. Historical parallels and analogies can be drawn from a number of events. The protests over hedgerow removal have echoes in protests over putting the hedges in during the Enclosures. Modern resistance to new technology such as GMOs has resonance with the rural unrest of the 1830s brought about by the introduction of the threshing machine. Efforts by agricultural interests to protect their economic position through the Corn Laws of the nineteenth century is analogous to late twentieth-century pleas by farmers for increased subsidy.

The rise and fall of the coal industry (see Box 2.1) illustrates how another primary industry has been linked to economic development and to environmental impacts. Mining left a huge physical legacy of pollution and waste which is still being cleared up.

Secondary industry

The rise of manufacturing industry was the distinguishing feature of the Industrial Revolution which led to the UK being for a short time the 'workshop of the world'. The symbolic smoking factory chimney represented not only industrial activity but also the pollution resulting from it. Traditional portrayals of the Industrial Revolution as the work of heroic inventors and engineers such as Darby, Watt, Arkwright, Stephenson and Brunel, obscure the fundamental changes in the

Box 2.1

King Coal and the Industrial Revolution

In 1800 most coal was used for domestic purposes. Water power was favoured for industrial use. 'As late as 1842, two thirds of all coal dug from British pits was burnt in the home', (Hall, 1981).

The industrial importance of coal came with developments from about 1850 providing new industrial markets. New machines started to be powered by coal and new technology meant deeper shafts could be sunk and better drainage and ventilation installed in mines.

The development of the railways is tied up with the development of the coal industry (see Box 2.2).

Coal provided raw material to the iron and particularly steel industry – steel was used for ship plates and railway lines – coal found new markets since it could be transported by ships and railways – Coal exports rose from 750,000 tons in 1857 to 13 million tons in 1873 (Hall, 1981) – fuel for ships and railways provided important markets for coal.

The low smoke output of South Wales steam coal was much prized by the Navy. This was important when specialized steel production and the development of munitions and explosives meant the range of guns was significantly increased.

For ten years after nationalization in 1947 coal could not satisfy the markets. After 1957 the coal industry began a decline from which it would never recover.

The passing of the Clean Air Acts from 1956 onwards cut down demand for domestic coal. The railways changed from steam engines to diesel and later electric-powered locomotives. The growth of oil as both a basic raw material and energy source, the advent of North Sea gas taking over from town gas and the decline of the UK steel industry as a major market for coking coal all combined to devastate the coal industry.

Oil overtook coal as the most important source of primary energy later in Britain than in any other industrialized country apart from the Soviet Union.

From the late 1950s the British Government followed an energy strategy which diversified primary energy requirements. Nuclear power and coal were preferred for electricity production until the mid-1980s when oil and gas started to be favoured.

The fate of the coal industry in the last quarter of the twentieth century has been linked with electricity production. The economic wisdom of protecting the coal industry and securing government-brokered contracts to supply the coal-burning power stations has been controversial in economic, political and environmental terms.

organization of business in this era. The chief legacies of the Industrial Revolution are, first, the factory system (Pawson, 1978; Perkin, 1989) which led the way to Fordist mass production. Second, the substitution of hand labour by machines resulted in productivity rises which led on to the modern developments in automation, roboticization and computerization. Third was the application of power to industrial processes, from water to steam to electricity. Fourth was the development of an interlocking industrial complex of suppliers, contractors, producers, assemblers, consultants and distributors (e.g. the twentieth-century car industry). Fifth, the development of the capitalist imperative for growth led to the search for markets, expanding trade and fostering consumerism. Sixth came the development of the corporate outlook separate from ownership, which also leads to the ethos of growth. Compared to these structural legacies, the physical legacies of waste tips, contaminated groundwater and derelict factories, while significant, are relatively less important.

The development of manufacturing has had varying impacts on the environment. The early small-scale metal-based industries resulted in considerable local pollution through air pollution and land contamination. Much of this was in the rural areas although these quickly became what we now know as brownfield sites as the tentacles of urbanization spread. Even what we regard as the heavy industries such as iron and steel, mining and the heavy chemical industries were originally carried out on a small scale. However, as the scale of operations grew, so did both the amounts of pollution and land despoliation and the area over which it took place. The way in which manufacturing developed was linked to the growth of the railways (see Box 2.2), which opened up new markets and provided a valuable market in themselves for iron and steel products. The railways similarly opened up new markets for agricultural goods allowing foodstuffs to be moved over greater distances to the cities.

Nineteenth-century manufacturing industries were essentially the processing of natural materials (foodstuffs, cotton, wool, other natural fibres, leather), metal fabrication or the direct abstraction of minerals. The twentieth century has seen the rapid development of industries based on chemical manufacturing leading to artificial compounds (dyes, fibres, plastics) with the second half of the century giving way to the production of petroleum-based compounds (plastics, phenols, etc.). This has been accompanied by oil largely replacing coal as a source of both the hydrocarbon building blocks on which many of the compounds are based

Box 2.2

Railways, the Industrial Revolution and the environment

The development of the railway network is seen as the epitome of the Industrial Revolution. The railway age had been born with the Stockton and Darlington railway in 1825 and quickly grew to the peak of railway mania in 1846. The demand for railway lines, railway rolling stock, locomotives and the rest of the infrastructure – everything from track ballast to platform barrows all stimulated industrial production. The widespread adoption of the Bessemer converter after 1850 meant large quantities of steel (as opposed to iron) at competitive prices were available and so the market for steel-based goods, most noticeably railway rails, increased. Hannah (1983) has pointed out the railways presented an interesting paradox. Although they were one of the few industries of the time which were organized in a monopolistic way, they promoted *laissez-faire* practices for the manufacturing industries by opening up new markets and by exposing most of the country to market forces for manufactured goods.

The implications of the railways were wide-ranging. They provided new markets for manufactured goods and allowed the transport of bulky raw materials. The expansion of the railways meant urban markets became accessible to rurally produced, largely agricultural, goods. This changed the nature of agricultural production, most notably in the production of milk (Bagwell, 1981). This effect was felt first in the hinterland of London but then later the conurbations of the North provided an important boost for the pastoral North and West eventually. The opening up of the countryside by the railways provided an artery for new ideas and for new marketing opportunities, although it also acted as a vein for labour to move from the country to the industrial conurbations, which was important when agricultural fortunes waned and widespread urban migration ensued

One of the greatest impacts of the railways came with the opening up of the American West (Paul, 1988). Many of the early American railways were built with British railway lines and provided a valuable export for the British steel industry particularly from South Wales. Due to the American coal industry still being in its infancy, the high transport costs and large distances involved, the American railways were based on wood-burning locomotives which had an impact on both the air quality and timber resources of the areas through which they passed.

and as the energy source required for the synthesis to take place. The twentieth century became the century of bakelite, rayon, nylon, DDT, cellophane and PVC. Increasingly, organic compounds (pesticides, pharmaceuticals, foodstuffs) have become commercially more important with chemical compounds which interact with living bodies in increasingly complex and sophisticated ways becoming ever more central to the chemical industry. These interactions include pharmaceuticals in the case of humans, growth hormones in the case of domesticated animals and pesticides in the case of insects and other pests. One of the great environmental concerns is when these interact with living organisms at which they are not targeted, e.g. DDT entering the food chain of raptors and other birds of prey in addition to insects as intended. Recent decades have seen the increasing use of genetic manipulation as a commercial process with chemical processes giving way to biological ones. The early decades of the twentieth century saw fundamental changes to the nature, layout and organization of factories and to the development of mass production techniques. The closing decades of the century were marked by increasing flexibility and of production schedules being organized on a world-wide, as opposed to plant-wide, scale.

Jeremy (1998) has noted that the major industrial innovations of the twentieth century have seven features in common. They are a function of invention or discovery of the last two hundred years, they came out of the advanced economies, mainly coming from the work of scientists in the twentieth century and most have held out the promise of economic and social benefit. In addition, most have been developed by large corporations and/or governments since nearly all have involved large capital investment and a surprising number have been initiated by either the two world wars or the Cold War. With a list of these features it is not hard to see why large companies have continued to dominate the industrial picture.

Following the Second World War manufacturing fell in love with technology. There was a belief that technology would enable the environmental problems to be fixed and that either it would prevent the creation of problems or provide the means to remedy any unforeseen problems. This is a contention at the centre of many of the interactions between business and the environment that we can see today.

By the end of the century in some advanced industries we have reached the point where the technology has taken over the industrial process. Perkin (1989) states:

> What is happening to industry in post-industrial society is what happened to agriculture during the Industrial Revolution. Agriculture, with the aid of fertilisers and machinery, became more, not less,

efficient, and with a diminishing work force was able to feed a majority instead of minority of non-agrarians. Now industry is becoming so efficient, with the aid of robotics and computers, that a small minority of the population are able to produce the consumer goods for the non-manufacturing majority.

Tertiary industries

The service sector has made a considerable contribution to the economy and certainly to employment patterns over the last two hundred years, and is not only a feature of the late twentieth century. What has changed is the way in which the tertiary sector is regarded. In the early nineteenth century small businesses and the lower orders of society dominated the service sector with large numbers in trade and in service.

What was and was not counted as 'employment' needs to be carefully analysed. Due to the rather unsophisticated way in which employment data were collected, there was an under-representation in the workforce of the large number of women in work and of family and dependent labour. Domestic service was a large employer. By 1870 Checkland (1982) states that 1.8 million were in service, more than had ever been employed in agriculture and the majority of these were women. Soon large employers started to emerge in the tertiary sector with the banking and finance sector and the railways becoming large employers. In 1907 of the twenty largest employers those in the tertiary sector included the GPO, one assurance company and twelve railway companies and only five were from the secondary sector. In 1992 the top twenty were headed by British Telecom, the Post Office and the British Railway Board, followed by a further ten from the tertiary sector including seven retailers and two banks. Of the five manufacturing concerns, three of the five came from the food and drink sector (Jeremy, 1998).

Care has to be taken in the interpretation of these statistics since throughout this period the boundaries between service and manufacture have been blurred. The railway companies should not be seen as merely falling into the service sector since they have traditionally manufactured their own rolling stock. Now traditional manufacturers see themselves as selling a whole service to customers. For example, IBM do not see themselves as manufacturers of computers but as providing customers with a whole range of information services. Services have not left much in the way of physical environmental legacies in the way that agriculture, coal mining or manufacturing have done. However, one might draw parallels with some past historical events in the tertiary sector and the present day, e.g. the present boom in Internet service companies have analogies with the Railway Mania of the 1840s and the South Sea Bubble of the early 1700s.

The birth and growth of environmentalism

The school of scientific and moral thought that has prevailed over this time has conditioned much of the thinking about the relationship between business and environmentalism. Judaeo-Christian thought believes that the Earth is there for man to use. This differs from other religions such as Buddhism, Jainism or the beliefs of the American Plains Indians, all of whom believed in the coexistence of man and nature. Science has been seen as a way of man having 'dominion over nature'. It was believed that science was God's way of giving man the opportunity to carry out the Almighty's work.

Not only have the natural sciences had an impact on the relationship between the environment and business but also economics has made an impact ever since the days of the original classical economists. Adam Smith maintained competition in the market economy would produce the most beneficial outcome for society in total. However, it is only recently that there have been serious attempts to include all the factors which can be considered to be of overall benefit to society into the economic equation (Turner *et al.*, 1994) and so it is only recently that environmental factors, to which it is difficult to attach monetary value, have been considered.

Marxist thinking did little to help the environmental cause. The early Marxists regarded the environment as something to be used to increase production, although they accepted the need to control and change the direction of science and technology. Their contribution was to consider who should control the means of production rather than consider what effect production might have on the environment. The drive to the proletariat controlling the means of production did not yield great beneficial results for environmental protection. The eastern bloc nations prior to the collapse of the old Soviet Union had a poor environmental record with the Chernobyl disaster of 1986 being perhaps an unfair but lasting reminder of that record.

The basic tenets of environmentalism were a response to the industrialization of the nineteenth century. Environmentalism is basically a nineteenth-century invention. It was the Industrial Revolution that provided the grit for the pearl of environmentalism. As soon as the Industrial Revolution got under way and the wealth it generated started to trickle down to more and more of the population, then some individuals and groups started to question where the excesses of industrialization might lead.

Blunden and Curry (1985) have identified three strands of the environmental movement during its early days. First were the Romantics whose prime concern was the preservation of God's glory but as the world got smaller felt the balance of nature was being disturbed. They saw no need to improve nature and resented any intrusion by urbanization into the rural beauty. The second environmental group were the scientists preserving nature as an outdoor laboratory. They searched for the secrets nature could reveal rather than merely admiring its aesthetic value. The third group were the Utilitarians who wished to conserve particular aspects of the environment as an exploitable resource. In many ways, individual preservation groups of today can see their roots in this movement. All three groups came from basically liberal traditions. Lowe and Goyder (1983) pointed out 'environmental concern was an integral part of the late Victorian intellectual reaction to many of the tenets of economic liberalism'. It became part of the reaction against *laissez-faire* commercial objectives along with a range of philosophical and political viewpoints.

For several reasons the proto-environmentalists were always going to have an uphill struggle to get their views accepted. The role of religious belief in maintaining social order at the beginning of the nineteenth century should not be under-estimated. There was an expectation that those in authority were Christian gentlemen. The Protestant work ethic ran through society although there was a clear distinction between the church goers – the farmers and industrialists – and the chapel goers – the farm and factory labourers. For centuries the land had been a manifestation of power. In feudal times the land and those on it had been there for the landlord to do with as he (and it was he) pleased. Later, as a sign of ostentatious power, landlords had the countryside landscaped as deer parks and formal gardens with no regard to the 'natural' landscape or the tenants who were displaced and relocated at a whim. So the move to mine, build factories and despoil with waste tips was seen as part of the continuing use of the land.

A second contributory factor worked against the effectiveness of the environmental movement. During the nineteenth century the numerous enterprises were small, localized and were owned by locally powerful figures who provided livelihoods for many local people. Restrictions on proprietors only came about when the activities led to a loss of commercial competitiveness for their neighbours, which is how the original air pollution legislation – the Alkali Acts – came into being or when accidents or industrial disease reached unacceptable levels.

There were organizations with a concern for the environment that emerged during these times. Some of these are still active today, reflecting a continuing concern for wildlife, the countryside and the national heritage. Perhaps most influential in the early days was the Commons, Open Spaces and Footpaths Preservation Society which was founded in 1865. Others included the Society for Prevention of Cruelty to Animals (later to become the RSPCA) in 1824, the Society for the Protection of Birds (now the RSPB) in 1889, the Selborne Society (more correctly the Selborne Society for the Protection of Birds, Plants and Pleasant Places) in 1885 and the National Trust for Places of Historic Interest or Natural Beauty (still the correct title but better known now as the National Trust) in 1895.

Throughout the period of our study we see social change resulting in the individual questioning the established order. We see attitudes to class, religion, authority and moral values all changing. The ways in which these changed alongside each other is fully described elsewhere (Perkin, 1981). The First World War gave independence to many, especially the young. This rejection of the established industrial order culminated in the General Strike of 1926. In this spirit the mass trespass movement of the 1930s should be seen as a social rebellion with political undertones and marks an environmental protest that had many of the characteristics of later conflicts. In this case the 'enemy' had been the aristocracy but in many later battles the 'enemy' was to be business.

Present environmental concern can be traced to the libertarianism movements of the 1960s. This was a time when the orthodoxy of the establishment was brought into question. It was at this time that a number of works which are still relevant to today were produced. Issues raised by Rachel Carson in *Silent Spring* in 1962, by Paul Ehrlich in the *Population Bomb* in 1972, by Meadows *et al.* in *The Limits to Growth* in 1972 and Schumacher in *Small is Beautiful* in 1973 are all still relevant today. However, the tales of doom and of resources running out within twenty years did not take place and these inaccuracies put back the environmental cause. Over the last thirty years knowledge has matured and fervour mellowed.

Environmentalism now is a much more diverse movement than it was in the 1960s. In addition to the long-established societies like the National Trust and the RSPB, the 1970s saw the growth of more radical groups like Greenpeace and Friends of the Earth. These were different because unlike the earlier groups, they sought not just to preserve nature but to change society to improve the environment and were not averse to using demonstrations, press campaigns, lobbying, petitions and stunts (such as obstructing whaling ships or sailing into nuclear test zones). As well as fighting local issues like nuclear reprocessing plants, they were explicitly international in outlook, reflecting the increased global concern for the environment. The late 1980s saw the growth of more localized grassroots groups who formed to protest about local issues such as road building. From these groups there developed more radical anarchistic 'disorganizations' such as Earth First! which seemed not to have conventional organizational structures or membership and was thus impossible for the authorities to understand or control.

Reclaim the Streets is a direct action movement which encourages local people to hold street parties to bring traffic to a halt to highlight the way freedom for pedestrians has been lost by excessive car use. 'Localism' (Orr, 1992) has developed as a reaction to increasing globalization. It advocates increasing self-sufficiency in production and consumption in limited local areas, for example, people living and working in the same locality rather than commuting. But it also stresses the consumption of, say, locally produced vegetables from a nearby farm in preference to beans from Guatemala or peas from Zimbabwe bought in an out-of-town superstore. Local exchange trading systems (LETS) are a good example of developed localism in action. LETS are a modern form of barter. Someone who may be an expert bicycle mechanic could exchange one hour of labour repairing bikes in exchange for someone else agreeing to clean their windows three times in two months. LETS thus foster community involvement and local sourcing of goods and services. In certain communities LETS have become a parallel form of currency. The key point about LETS is that there is an agreed local system of equating the value of different goods and services so that a certain number of LETS tokens are always exchangable for the same thing, e.g. a box of vegetables and four hours of babysitting may both be worth the same number of LETS tokens.

Environmentalism has not had a completely unchallenged progression in influencing business and government. There has been what Rowell (1996) calls the 'Green backlash'. There has been a reaction from certain businesses and other groups who oppose the actions of environmentalism. A major example is the Global Climate Coalition (GCC). This is an association of mainly energy producers and other businesses dependent on fossil fuels that feel threatened by proposals to curb carbon emissions. If governments did legislate internationally to reduce greenhouse gas emissions, energy industries would stand to lose profitability. The name of the group sounds neutral or even pro-environmental, but its aims are explicitly against regulation of energy and to counter environmentalist arguments about global warming. The GCC website (http://www.globalclimate.org/mission.htm) highlights any reports which cast doubt on global warming being caused by industrial activity.

Business opposition to environmentalism is not always so obvious and the 'hijacking' of the environmental agenda can take place when corporations appear to accept environmental ideas, but do so in a half-hearted or deliberately ineffective way simply to deflect criticism (Welford, 1997). Business associations such as the European Round Table of Industrialists representing forty-seven of Europe's largest industrial companies (ERT, 2000, online) pursues a similar agenda to the GCC. Suspicion of the business world's motives led to the setting up of Corporate Watch, an organization that monitors the activities of businesses and publishes its findings on the Internet (Corporate Watch http://www.corpwatch.org).

Environmentalism itself had become globalized by the end of the twentieth century. The Internet was harnessed to mobilize an international coalition of protesters at the WTO millennium round meeting at Seattle in November 1999. The talks ground to a halt when thousands of environmentalists, trade unionists, anarchists, pacifists, human rights activists and Third World farmers protested on the streets against WTO proposals for further trade liberalization while delegates from developing countries refused to accept the proposals of the rich countries. Scenes of dramatic confrontation between protesters and heavily armoured riot police were broadcast world-wide (George, 2000). The concerns of environmentalism are now much wider than they were in the past and are meshed in hybrid issues together with social justice, democracy and human rights. Both business and environmentalism find themselves in a much more complex world than that of thirty years ago.

Summary points

This chapter has drawn on a number of academic disciplines to show the way in which the various elements that make up the central theme of the book are closely related. The themes which are considered are:

- the physical environment;
- business processes;
- industry;
- environmentalism.

All of these have been subject to considerable change or evolution over the last two hundred years (see Table 2.2, pp. 56–7). The chapter shows that environmentalism and modern business both had their roots in the Industrial Revolution and the four factors have been closely associated ever since. Over the period there have been marked changes in the development of each but this has triggered changes in at least one of the other developments. After the Second World War there was a period of technological change. The new industrial order stimulated a new way to do business. These two factors together started to have an impact on the environment and as the wealth trickled down environmental concern grew. We have attempted to show relationships rather than causation. We have merely pointed out how the various strands have developed together. By looking at the changes through time we can see the way the various elements have developed together.

Further reading

Further reading for this chapter falls into three very distinct areas from three contrasting disciplines. For an understanding of the way in which the physical environment has changed over the two hundred year period a number of geographical textbooks are available. A fairly complete coverage can be achieved by reference to I.G. Simmons, *Changing the Face of the Earth* (Oxford: Blackwell, 1996) which presents a good coverage of those areas of relevance to this text. For specific cases of environmental processes see A.S. Goudie, *The Human Impact on the Natural Environment* (Oxford: Blackwell, 2000) and, by the same author, *The Human Impact Reader* (Oxford: Blackwell, 1997) which is a collection of thirty-nine separate contributions – just about all highly relevant as well as a table of key textbooks relating to the human impact on the environment. The approach taken by C. Ponting, *A Green History of the World* (London: Sinclair Stevenson, 1991) provides a valuable contrasting viewpoint.

Much use has been made of the historical literature. R.A. Dodgshon and R.A. Butlin, *An Historical Geography of England and Wales* (London: Academic Press, 1978) has proved of great value and if going back only two hundred years is not far enough, this is an excellent starting point. H.J. Perkin, *The Rise of Professional Society* (London: Routledge, 1989) along with G.E. Mingay, *A Social History of the English Countryside*, (London: Routledge, 1990) explains the social changes of the last century while D.J. Jeremy, *A Business History of Britain 1900–1990s* (Oxford: Oxford University Press, 1998) provides a full account of the evolution of business practices this century, although many readers may be more familiar with S. Pollard, *The Development of the British Economy 1914–1990* (London: Edward Arnold, 1992). For a detailed examination of agricultural change, D. Grigg, *English Agriculture: An Historical Perspective* (Oxford: Blackwell, 1989) provides a comprehensive account which extends through the twentieth century and is probably more digestible to the average reader than the classic work of E. Kerridge, *The Agricultural Revolution* (London: George Allen and Unwin, 1967). J. Blunden and N. Curry, *The Changing Countryside* (London: Croom Helm, 1985) paints a clear picture of the changing countryside particularly in the early part of this century.

The way in which environmentalism has evolved is more a matter of political viewpoint than of indisputable fact. However, the text by T. O'Riordan and R.K. Turner, *An Annotated Reader in Environmental Planning and Management* (Oxford: Pergamon, 1983) provides valuable sources of information. No understanding of the way in which the countryside has evolved would be complete without reference to O. Rackham, *The History of the Countryside* (London: Dent, 1986) and D. Evans, *A History of Nature Conservation in Britain* (London: Routledge, 1997). Neither has been used extensively in this text but both provide valuable contributions.

Discussion points

1 How has the organization of business changed over the last two centuries and what impact has this had on the environment?

2 Compare the reasons for the rise of environmental groups in the nineteenth century with those of the late twentieth century.

Exercise

Compare a modern 1:50,000 sheet map of any industrial area with the nearest equivalent map that you can find for the nineteenth century (a reprint map is usually available for most areas). In what ways has the landscape changed in terms of settlement, transport, industry, mining, etc.? What has happened to woodland, water and other open land? Make some sample estimates of changing percentages of broad categories of land use. If you have time, visit the area with the two maps and see what is the present situation.

Table 2.2 ***Comparative developments in environment, business, agriculture, industry and environmentalism over time***

	pre-1800	*1800–1900*	*1900–1960*	*1960–2000*
environment	much natural habitat in existence; little pollution, few cities; landscape in transition	urbanization on bigger scale; loss of habitat; increased waste and pollution; national integration of space and time	urban sprawl; suburbanization; decline of natural habitats; decline of species; local pollution problems, e.g. urban air pollution	counter-urbanization; resource depletion; increasing waste and pollution; global environmental issues, e.g. changing climate, biodiversity
business	early capitalism innovators and entrepreneurs; Protestant work ethic; small local business; Adam Smith	*laissez-faire* capitalism; mostly small-scale but large firms developing; limited liability; modern business ideas developing; Karl Marx	diversification of activities; increasing regulation; growth of corporations; government intervention J.M. Keynes	deindustrialization; service sector growth; multi-national firms; deregulation, trade liberalization; globalization; Milton Friedmann
agriculture	local production; dominant type of work, era of Agricultural Revolution	Enclosure; Corn Laws; high farming; beginnings of flight from the land; start of mechanization; Great Depression in agriculture	wartime protectionism; mechanization and chemicalization; vast productivity rise; post-war protection and subsidy; decline in employment	big decline in farms; increased farm size; high cost/high output; over-production crisis; low world prices; BSE and GMO crises; pluriactivity

Table 2.2 ***continued***

	pre-1800	*1800–1900*	*1900–1960*	*1960–2000*
industry	small scale; labour intensive; factory system beginning; water power, canals; iron; textiles	steam age; steel/coal; machinery increasing; railways, steamships, telegraph	oil age; electricity; chemicals, automobiles; electrical goods; consumer goods; telephone/radio; Fordism	nuclear power; oil/gas; television; computers; internet; biotechnology; renewable energy; consumer goods; post-Fordism
environmentalism	subdue the Earth; mastery over Nature; improve natural world; landscape parks Gilbert White	Romantics; Natural History; first environmental groups, e.g. RSPB, NT; Charles Darwin	recreation and amenity groups, e.g. CPRE; Ramblers Association; Ecology; scientific approach grows; government legislation on wildlife	environment crisis; mass media interest; radical pressure group growth, increasing legislation; more agencies; public concern grows; environment becomes a political issue

3 Environmental business perspectives: assets, costs and externalities

- How business works – internal and external relationships
- The concept of environmental assets
- Environmental business costs
- Public environmental costs
- The importance of environmental externalities for business
- Subsidies and compensation
- Transformations
- Further factors causing friction
- Measuring and valuing the environment

> Pressure from consumers is not sufficiently universal or coherent to ensure that dirty firms are driven out of business and environmentally responsible ones prosper.
>
> (Caincross, 1995)

Let us begin by considering a commercial enterprise. It may be a farm, a chair manufacturer or a public house – examples picked since they represent the primary, secondary and tertiary sectors of the economy. To carry on the business the farmer needs tractors, ploughs and barns. The carpenter who makes chairs needs a workshop, a lathe and various tools to make the chairs. The publican needs barrels of beer, glasses, tables and a public house. These are the *assets* of the various firms. An asset is anything (whether material such as property, buildings, machinery, etc. or intangible such as an idea or reputation) owned by a company which is used now, or which may be used in the future, to assist in the conduct of the business. Even if an asset is not used, it can always be sold or traded for something more useful. Hence buildings may be sold to raise cash. It is usually necessary to have assets in order to do business.

Assets are normally categorized as current assets (cash, current account sums which can be readily liquidated, stock, work in progress) or fixed assets (land, buildings, plant or machinery). To this might be added intangible assets (patents, intellectual property, goodwill, reputation), although putting a value on these may prove difficult.

To carry out their business our farmer, carpenter or publican incurs expenditure. The farmer has to buy seed, fertilizer, diesel, employ labourers and pay his accountant. The carpenter has to buy wood, paint and screws. He has to pay for advertisements if he is ever to sell any chairs. He has to heat and light his workshop. The publican has to employ bar staff, has to buy the beer from the brewery and has to pay for the pub to be painted and decorated regularly. These are all the *costs* of running a business. A cost is anything that must be paid out in order to carry out the business. They are the necessary (and sometimes unnecessary) elements of a business operation. They are fully recognized, accounted for and paid for by the business. Costs are the price a business expects to pay to achieve its objectives.

All of our businesses, in addition, use things for which they do not have to pay directly and so do not have to account for them in their pricing structure. The farmer receives free weather information. He does not have to pay his own weather forecaster – the state pays for the Meteorological Office. The carpenter needs training both in his craft and to run his business. Although he may pay a contribution to any training, most education and training is still provided by the state. The publican's business results in considerable noise and disturbance to those living close to the public house. This means neighbours have to either buy ear plugs or pay to have their properties double glazed. The publican is unlikely to have to pay for such remedies. He does not have to allow for them in the price he charges for his beer. The costs are borne by others. He is said to have externalized his costs.

It is easy for us to appreciate the concept of assets and costs. *Externalities* are slightly harder to understand. However, we shall see that they are very often associated with the impact of business on the environment. Many externalities involve the environment and constitute an important concept in both environmental and welfare economics.

Environmental costs, assets and externalities are simply one class of general costs, assets and externalities. Table 3.1 shows a typology of costs, assets and externalities, according to whether they are intrinsic to the business (i.e. cannot be avoided as they are essential to its functioning), imposed by society (in the form of regulation or societal pressure) or are capable of being transformed from one type to another.

Table 3.2 lists a selection of assets, costs and externalities. This is not intended to be an encyclopaedic list but to indicate the very wide range of items possible. Some elements may in fact appear in any column, but for clarity we have confined items to their most usual category.

Table 3.1 ***A typology of costs, assets and externalities relating to the environment***

	Costs	*Assets*	*Externalities*
intrinsic to business	unavoidable and necessary for business operations	necessary for present business operations	unavoidable consequence of business
	necessary for future development	necessary for future use	avoidable but uncontrolled, no pressure to change
	unnecessary costs due to inefficiency	unrealized assets – foregone opportunities	capable of being internalized – pressure for change
imposed by society	costs of compliance to environmental regulations	new opportunity created by changing social behaviour or tastes	limits to environmentally unacceptable practices
	costs of forestalling legislation	environmental legislation forcing re-evaluation of assets	planning framework sets constraints
	costs to image – loss of custom, fall in share value		social acceptability
	costs of disobeying laws – fines		
capable of transformation	cost to asset (benefit) cost to externality (benefit)	asset to cost (liability) asset to externality (benefit or cost)	externality to cost (liability) externality to asset (benefit)

Table 3.2 ***Examples of environmental assets, costs and externalities***

Assets	*Costs*	*Externalities*
land for a site	energy	air pollution
clean water	insulation	radiation
valued landscape	air conditioning	chemical pollution
heritage site	pollution control equipment	dust
historic building	containment building	smoke
ancient woodland	protective clothing	grit
forestry plantation	modified process	water pollution
fertile soil	fines	smells
wildlife	taxes	noise
mineral resource	costs of compliance	vibration
viewpoint	inefficiency	spatial severance
country house	compensation payments	light pollution
climate	legal costs	thermal pollution
accessibility	waste disposal	visual intrusion
production subsidy	landfill	waste
environmental subsidy	transport	resource erosion
tax concession	insurance	land use incompatibility
image/reputation	forestalling legislation	social pathology (crime)
recycled materials	advertising/PR	land value decline

How business works – internal and external relationships

In simple terms a business uses its assets (land, labour, materials, ideas, reputation) to produce and sell some material good or a service (Figure 3.1). In the process of doing this, the business inevitably incurs costs, such as rent for land, wages for workers, charges for power and water, taxation on profits, etc. The firm has to raise its assets, which are necessary for its operations. It does this by either seeking loans, partners in the business or raising share capital. It also has to pay for those costs that are a necessary part of its operations. It does this by making profits. It makes its profits by selling goods and services at a sufficiently high price and keeping its costs in check. If it is to expand, it must ensure it has secured sufficient assets to make all this still possible and that the cash flow going through the business is sufficient to support its operations. A business which does not generate enough profits to do all these things will eventually fail. Business is therefore necessarily profit-orientated and growth-orientated. It is built into the culture of capitalism

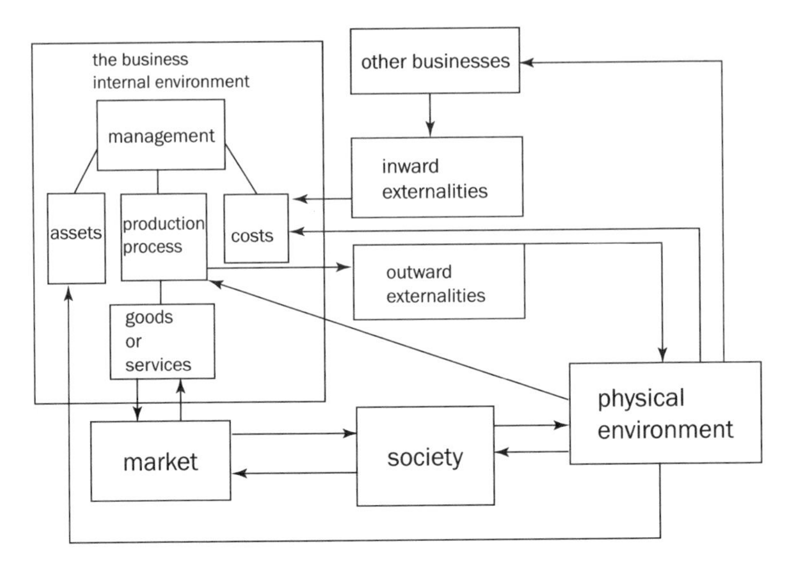

Figure 3.1 ***Relationship between the internal and external business environment***

to make profits and to seek growth. To increase profits, part of the strategy of business may be to increase market share, or to decrease costs by increased efficiency. This may lead to a business exporting a cost rather than bearing it internally. Somebody else will then have to meet the cost. However, the real world is imperfect. Knowledge is not complete and people are fallible. As a result, businesses do not always succeed in reducing their costs as far as they would like, or may not even be aware that their costs could be reduced. Some part of the potential cost reductions may be attributable to under-appreciated environmental costs. Howes *et al.* (1997) highlight a survey of businesses involved in a waste prevention programme that under-estimated by 90 per cent the true costs of their wastes. There was genuine surprise and not a little embarrassment among executives that they had not previously identified the true costs of their waste. It should be remembered these were businesses who were interested enough in the environment to voluntarily take part in a waste reduction programme. This begs the question, what is the cost of waste to the majority of disinterested businesses?

The concept of environmental assets

Environmental assets are useful attributes of the environment owned by a company for use in its business, for example, land, raw materials, or a landscape. An asset should be financially measurable in some way. It should be possible to sell it, to use it and add value to it, or to use it and lose value or to do nothing with it and let it appreciate in capital value. The idea of positive utility is central to the idea of an asset. An asset is something valuable that is desirable to have. It can be valued and used to create wealth. Many aspects of the environment have multiple values as assets and may be used in a variety of ways. As an example, consider the case of a piece of woodland owned by a company. It may be:

1 a source for a product:
   (a) for sale as a raw material, e.g. timber;
   (b) for manufacturing to add value, e.g. timber made into furniture.
2 a source for a service, e.g. a wood used for recreation for which admission is charged.
3 a financial asset:
   (a) land for future sale at higher value;
   (b) as security for a loan or mortgage;
   (c) for tax advantage;
   (d) for production subsidies;
   (e) for environmental subsidies and grants.
4 as a marketing opportunity:
   (a) used in advertising products/services;
   (b) used as public relations-image building 'we really care about the environment';
   (c) used for place promotion-photograph on a brochure.

An asset is used to further business aims but it may also be used to further environmental objectives. An environmental asset may have multiple uses, being used either consecutively, alternatively or simultaneously. So, for example, the woodland could be used solely as a source of timber, or it could be used solely for hunting, or solely for bird watching, as nature reserve, campsite or as a site for paint ball games. Or it could be used for a combination of several of these functions (though not all at once!). A company primarily concerned with timber extraction might set aside a part of the wood for nature conservation for which it would attract financial benefits from the state in the form of subsidies and grants. Such a move might also receive public relations benefits by creating an image of caring about the environment. Such an initiative may be genuine, or simply 'green wash' intended as a cosmetic sop to deflect any deep criticism of the company's main activities. The real reason will depend on the extent of the reserve, how effectively it is managed, how secure its future is in the long term, how far it is affected by the commercial activities elsewhere in the wood, its intrinsic ecological value and what sort of access is available.

Environmental assets can change into assets from artefacts that had no intrinsic financial value. For example, Stonehenge was 'there' for thousands of years. However, it is only recently that it has become the much visited 'heritage attraction'. It now charges for admission and provides added value in the form of souvenirs, bookshops, café, toilet and interpretation via information boards. Some 58 per cent of British residents say that they would be willing to pay up to £7 extra in taxes over two years to protect Stonehenge (Kennedy, 1999, p. 10). Such is its value as a cultural icon that people are willing to pay for its upkeep even if they receive no direct benefit from actually visiting it. A business asset only exists once someone perceives it to be a feasible commercial opportunity. A marked trend of the last quarter of a century has been the steady commercialization of aspects of the environment that were formerly free. An interesting example of how business can capitalize on an unrealized asset is the formation of a Welsh farmers' film location agency. The agency finds suitable film locations for movie makers, television and advertisers on a variety of landscapes around Wales. Farmers are therefore cashing in on their landscape assets directly (Kennedy, 1998b). The motivation may be simple business profits or it may be more complex as with environmental organizations seeking primarily to safeguard the integrity of a site but needing to adopt business methods in order to raise finance; a theme we will be exploring later. It becomes debatable how far an environmental organization can go before its fund-raising activities become indistinguishable from a profit motivated business (see also Chapter 9).

Environmental assets can be classified in a variety of ways. Owen (1991) divides such assets into three types:

1 critical natural capital: for example, the ozone layer, the tropical rainforest and the oceans. They cannot be damaged without extremely adverse consequences to the whole biosphere let alone the economy. Such assets are generally not owned by any one person, company or nation and are therefore 'free goods'. Despite their criticality, they can be, and often are, abused by business activities to the detriment of all. Since they are not owned, they are outside conventional accounting schemes.

2 other natural capital: for example, mineral resources or agricultural land. These assets are normally owned by a business, but they may be subject to depletion as they are incompletely valued in environmental terms.
3 man-made assets: for example, infrastructure and plant. These are clearly owned and are created purposefully by business and are usually properly valued by conventional accounting methods.

The first type of asset is often not properly accounted for at all and is regarded as an externality (that is something outside the company's control and responsibility). It may or may not have an impact on the firm one way or another. In addition, it may be under-valued, since only its strict monetary value is assessed. For example, the value of the Pyramids would be assessed in terms of the building value of the stones and real estate with no value being allocated to their historic associations. The second and third type of assets fit more easily into conventional accounting methods although the wider environmental values of both are often ignored. For example, agricultural land has an aesthetic value as a landscape as well as a monetary one for growing crops. Factories may be valued for their role in the production process but their costs as polluters of the environment are not usually taken into account. A typical business view is that: 'ecological issues are long term in nature and particularly affect future generations, whilst environmental impacts are externalities as far as the entity itself is concerned' (Owen, 1991, p. 24).

Environmental assets are intrinsic to the business for present operations, for example, a superstore needs a site upon which to trade. Other assets may not be necessary now but are kept for future operations. A mineral company may hold the lease on ore-bearing land but not plan to mine for ten or more years. A type of hidden asset is an unrealized one. In other words, the company possesses an asset but does not use it and thus foregoes an opportunity, e.g. a timber company has a forest that possesses valuable salmon fisheries, but does not rent them out. Assets can be imposed by society through changes in social tastes or behaviour. The upsurge in interest in heritage has created a huge demand for visits to country houses or ancient monuments. What was once visited freely by a few enthusiasts is now a mass recreational experience and has become commodified. A ruinous collection of old industrial buildings can be rebranded as a heritage theme park and becomes an asset. Environmental legislation may force a revaluation of assets previously ignored, e.g. laws to curb sawdust pollution from sawmills may prove the stimulus to use the sawdust as a basis for a fibre board industry. So the disposal of sawdust which was once a cost has now been turned into a source of revenue.

Environmental business costs

Costs are intrinsic to doing business and some costs cannot be avoided. They are part of business operations and are included in the price of the finished goods and services. You cannot produce bread without flour, for example. The costs of using this raw material are accepted as part of the firm's cost structure. Costs may be also running costs. These are necessary for the present production operations (e.g. electricity used to power machinery) or they may be capital costs (such as cost of the land for the factory site). Some costs may be necessary for future developments where a firm pays for an asset now for future use (e.g. land for a forestry plantation that will not be planted for five years). Further costs may be due to inefficiency (heat loss due to poor insulation).

In addition to its general operating costs firms incur a series of costs relating to the environment. To some firms these are not particularly important, to others they are crucial and can account for a significant proportion of their overall costs. In all cases, these costs are a legitimate and justifiable part of its operations for which it must pay. As with all other costs it makes sense to try to reduce these by as much as possible and to avoid them altogether if at all possible. Firms who internalize externalities become liable to an extra cost. However, it may be in a firm's long-term interest to internalize some environmental costs even if in the short term these could be externalized. This may be done for a variety of motives, including compliance to regulation, anticipation of future regulation, image building or eco-efficiency. Some environmental costs may relate directly to the processes involved in the product or service, for example, the purification of water supplies to the required level or the cost of fertilizing agricultural land to ensure the required yield of crop. In addition to what might be termed direct environmental costs are a series of other costs which are imposed by society but which have a strong environmental dimension. These include the cost of compliance with environmental regulations which include pollution emissions limits (e.g. having to install desulphurization equipment in order to burn coal). The cost of compliance may well involve costs in the form of control equipment, monitoring, extra staff, etc. and is accepted in the same way as health and safety and labour legislation. Planning laws may also impose costs on business by influencing the location of the business, perhaps to a higher cost site than might otherwise be the case, or by stipulating environmental conditions such as tree planting, before planning permission is given. While compliance is necessary if firms are to stay within the law, it also brings with it hidden benefits. These include energy savings and the potential for a better public image with the possibility of enhanced sales. If firms do not comply, then the sanction is that they are subject to fines, legal costs or sometimes ultimately imprisonment. The level of fines needs to be such that firms realize it is in their best interests to comply with the legislation.

Another cost is the cost of complying with social pressures, trends and mores – if public opinion is sufficiently strong, which may occur if issues have been taken up by environmental NGOs and the media, it may be necessary to incur costs to meet the objections. For example, Shell backed down over the disposal of the Brent Spar oil platform due to pressure group and media activities, which had led to a consumer boycott in Germany. As a result the company experienced higher costs in towing the platform to a Norwegian fjord (Box 3.1).

Where firms fail to pick up on social trends it may involve additional advertising revenue to counter an image of environmental unfriendliness. Improving the image or getting the environmental image over can involve considerable costs. Even worse, failure to respond to public concern may result in a loss of confidence by consumers, a loss of market share and a fall in share price. The situation can be made worse if competitors are seen to be making active steps to improve their environmental behaviour at the same time.

Costs involved in improving the environmental image of the firm with the aim of increasing profits or market share can be significant. DIY chains have boycotted unsustainable tropical timbers and stocked only sustainable timber to improve their image. Sometimes firms will be forced to use more expensive raw materials since cheaper sources are environmentally unacceptable. Esso's £3 million sponsorship of the Millennium Tree planting project in the UK or Monsanto's £1m newspaper advertising campaign in August 1998 to promote the idea of genetically engineered food crops to consumers are further examples of attempts to improve a firm's environmental image (Box 3.2).

Firms may choose to incur costs not required by legislation in order to forestall stricter legislation in the future, i.e. to set the agenda themselves rather than let the EU or the national government do it since this might

Box 3.1

Brent Spar

> Oil rigs are amazing pieces of modern technology when they are first built and potentially dangerous hunks of junk at the end of their lives. The dilemma of what to do with them is typified by the storage buoy Brent Spar. The International rules are clear, everything in water less than 75 metres deep must be removed – and that means most things in the North Sea.
>
> Shell (1998)

The former asset of Brent Spar became an increasing liability for Shell when they decided to remove it from the North Sea and dump it in deep water in the Atlantic in 1995. Environmental groups, particularly Greenpeace, mounted a major campaign to get Shell to back down and remove the buoy altogether from the sea, since they claimed that Brent Spar would prove to be a major environmental hazard to the Atlantic marine ecosystem as it gradually decayed. A sophisticated media campaign with Greenpeace providing the video footage for TV news broadcasts cast Shell as the villain and a considerable consumer backlash built up, particularly in Germany where boycotts of Shell petrol were organized. Despite public support from the then Prime Minister John Major in the House of Commons, Shell decided to accede to the pressure and Brent Spar was towed back and eventually ended up in a Norwegian fjord for final decommissioning. The 'hunk of junk' proved to be an acute embarrassment to Shell and its costs in terms of poor publicity was more than the actual monetary costs of disposal.

Sources: www.Shell.com. Encyclopaedia Britannica: Book of the Year (1996), 'Brent Spar'.

result in tougher and costlier measures. For example, chemical companies voluntarily have phased out CFCs (Eden, 1996). Costs may result from expenditure on lobbying and legal fees in order to persuade legislators not to introduce potentially costly regulations (e.g. setting up a trade organization to produce guidelines and voluntary codes of conduct, see Eden, 1996). In the case of voluntary costs, the business is accepting increased costs in the short term, believing that it will bring long-term benefits through improved image, greater market share or more favourable legislation.

Insurance costs, either from a commercial insurance company or self-insurance against liability due to hazardous processes or operations are

Box 3.2

Monsanto and genetically modified organisms

Biotechnology company Monsanto attempted to persuade UK consumers that genetically modified organisms (GMOs) were perfectly safe and indeed beneficial to health and the environment through an extensive £1m advertising campaign in August 1998 and through its website (www.monsanto.co.uk). A typical magazine advertisement showed an ordinary-looking potato and asked: 'How does this potato differ from a potato? This is a potato which has been grown using plant biotechnology. A naturally occurring beneficial gene has been inserted into the genetic structure which makes it insect resistant. It looks like any other potato. It does not taste any different and as a result of biotechnology it was grown using 40 per cent less chemical insecticide.'

Monsanto invited readers to look for more detail at its website where GMOs were described in more detail and the benefits of biotechnology explored. The company also gave web addresses of opponents such as Greenpeace and the Natural Law Party and invited readers to make up their own minds. However, GMOs suffered a good deal of adverse publicity in the autumn of 1998 when English Nature, the government's wildlife adviser, recommended a moratorium on crop trials of GMOs, the Vegetarian Society decided not to endorse products containing GMOs, Iceland Frozen Foods decided to ban GMO products, as well as direct action by eco-radicals who uprooted crops grown in trials. The whole area had become so sensitive that the October issue of *The Ecologist* dealing with GMOs was pulped by its printers who feared becoming embroiled in litigation despite both the editor and Monsanto declaring that they could not understand why the edition had been withdrawn.

Farmer's Weekly, Britain's leading farming periodical, sponsored a conference entitled 'Genetic Modification – path to profit or road to ruin' in February 1999, the title of which sums up many of the arguments about GMOs. Are they the answer to improved profitable farming or are they an environmental disaster in waiting? The government response was to announce strict controls on their cultivation and monitoring of their environmental impact on 21 October 1998. Environment Secretary Michael Meacher said: 'I feel strongly that the use of GM crops in agriculture must not put unacceptable pressure on our countryside and wildlife, and prejudice our goal of maintaining and where possible enhancing farmland biodiversity.'

D. Parr, Greenpeace campaign Director, replied: 'The government has been driven to respond to wide public antipathy on genetically modified organism (GMOs) but the biotechnology industry is still dictating the terms.'

The GMO controversy is a good example of how a MNC (sales of $10.7bn and market capitalization of $22 billion) has to take account of public concern and is prepared to encounter considerable costs in its strategy for technical innovation acceptability. By the autumn of 1999 Monsanto conceded that its publicity campaign had largely failed to persuade UK consumers.

Sources: Monsanto (1998); *Farmers Weekly* 23 October 1998, p. 6, p. 18; Garrett (1998); Greenpeace (1998a).

important. Sometimes there are costs in the form of higher premiums where environmental hazards, such as more frequent drought or floods, are likely to increase insurance claims.

Relocation to avoid unacceptable costs: sometimes companies will relocate to by-pass strict environmental regulation in their own country and move to developing countries which have limited or non-existent environmental regulations. They aim to reduce costs and then export the finished product back to their home country. Sometimes this can have tragic results as with the explosion at the Union Carbide plant at Bhophal in 1984. Environmental influences can lead to the search for new markets for products banned in the home countries, but which are still unregulated in the Third World; for example, DDT was banned in most Western countries as environmentally hazardous but was still exported to poor countries who did not control it.

Public environmental costs

It is difficult to estimate the true cost of the environment to a nation, since only those things with a price tag are usually included. The DOE (Brown, 1992) admits its figures are estimates and should be regarded as 'broad orders of magnitude only' (p. 237). Table 3.3 shows a comparison of the costs of the environment in the UK and Germany for 1990.

Table 3.3 ***Public costs – expenditure on the environment in the UK and Germany, 1990***

	UK	*Germany*
total £	14 billion	35 billion
% GDP in	2.5	6
% spent on		
air	16	47.3
water	47	17.2
waste	21	n.a.
noise	4	3.8
land	7	5.8
other	5	25.9

Source: DOE (1992); Lester (1992)

The UK figures represent actual expenditure while the German figures refer to estimated costs and include additional items to their UK counterparts, essentially more intangible factors such as the dangers to health, loss of residential value and loss of productivity. The similarities and differences are revealing. Bearing in mind that Germany is a larger country than the UK and that the basis of measuring is different, the UK estimates that the environment costs 2.5 per cent of GDP while Germany estimates the figure to be 6 per cent. The Germans appear to spend a far higher proportion on other elements (which are taken to include the separate waste category of the UK). Although imperfect, the data indicate that some aspects of environmental costs can be measured quantitatively, although they appear to be underestimated in the UK system of measurement.

Table 3.4 ***Expenditure on the environment by different groups in the UK, 1990–91***

	£ millions	*%*
government	4,800	33
enterprise	8,700	61
households	680	5
non-profit-making organizations	160	1
total	14,000	100

Source: DOE 1992, Table 17.2

Who pays for the environment is always going to be a contentious issue. It seems from Table 3.4 that business is already paying the biggest share (61 per cent) and so can claim with some justification they are already doing their bit for the environment. Government, by contrast, is some way behind but still the second biggest contributor with 33 per cent. Households appear to contribute a mere 5 per cent, but since the government funds ultimately come from taxation households are paying indirectly more than at first appears. It can be argued that some of business's contribution is passed onto customers and so householders are in fact also paying some of the business contribution.

The importance of environmental externalities for business

Environmental externalities occur where there is some aspect of the environment for which the business does not pay but where there is a cost imposed on somebody else – very often society at large. We shall see that while assets and costs are to be found in all aspects of business, externalities are particularly associated with environmental concerns. They are a major topic in environmental economics, which is concerned with how externalities are measured and how best to resolve the dilemmas they throw up. A major reason why externalities are of particular concern to the interaction with the physical environment is that individuals, companies or the state do not own many of the elements. In the absence of clearly defined property rights the charging of these elements presents a particular problem. Firms have been able to regard the air, rivers, wildscapes and oceans as free goods. Since it is clearly not in their interests to increase their own private costs, it has usually been the pattern that these costs are externalized since, by definition, firms have chosen not to incur them as private costs. The market traditionally has had no mechanism for apportionment of such costs to their rightful source, for example, no one 'owns' the air and so when a haulage company creates pollution from the operation of its vehicles, it does not pay for the right to pump out exhaust gases. It has created an external cost that will be borne by the people who choke on the fumes. Most forms of pollution are externalities. These can be regarded as either:

(a) extraneous costs which have been imposed from outside the business by external events or organizations, e.g. other businesses may impose an external cost on the business by having polluted water upstream which must be cleaned before it can be used. This is therefore an extra imposed cost.
(b) costs which a business passes on to society, e.g. air pollution from a factory. Other businesses have to filter the air; people suffer bronchial complaints and the ecosystem suffers from increased toxicity. The perpetrator has in effect passed on the costs to others.

So the polluters are able to reduce the costs of their operations, by passing on the pollution, instead of incorporating it into their cost structure by dealing with it at source. Recognition that externalities are paid for by someone and that equity demands that the 'polluter pays' are at the heart of much of our current environmental regulation. It is perhaps worth noting that externalities can be beneficial as well as detrimental. So a café next to a major heritage site receives the benefit of an unexpected flow of visitors passing its front door. It has done nothing to attract this resource and does not have to pay for it although it receives the benefit. Education and training are other beneficial externalities since if a firm employs a skilled workforce they receive the benefits that those staff bring with them without incurring any of the cost for how those skills were acquired.

Traditionally externalities have been off-loaded onto the wider community and the company creating them has escaped having to make any payment. In the past, companies would claim that externalities were unavoidable. Many externalities are avoidable but at present uncontrolled because there is no stimulus through regulation or consumer pressure to internalize them. Externalities arising from someone else's actions are frequently part of a company's costs, e.g. congestion costs imposed on deliveries due to heavy traffic from elsewhere. In theory, most (if not all) externalities could be internalized through new technology, legislation or changes in consumer pressure. While they could be internalized through altruistic actions by polluters, this course seems unlikely. Society imposes limits to environmentally unacceptable practices and has done in some form for as long as people have lived in organized groups, for example, in Imperial Rome, in order to reduce noise and congestion, there were restrictions on the hours when heavy wheeled carts could enter the city. In the thirteenth century, Edward I of England famously introduced the death sentence for anyone burning coal in London (there was at least one unfortunate victim of this policy) (Elsom, 1992).

In more recent times it has been appreciated there is an injustice in allowing such externalities to go unchecked, since it exerts a high price on the environment. Since the 1920s economists, planners and legislators have made some headway in determining instruments by which environmental costs can be correctly and efficiently allocated. This means the externalities can be internalized – a concept at the heart of the 'polluter pays principle'.

Externalities are now limited through the operation of the planning system, e.g. by zoning land uses and forbidding certain activities within particular zones. In addition, society imposes agreed standards of behaviour. If any company chose to ignore these they would lose business long before the law caught up with them; e.g. a restaurant which dumped its refuse in a midden outside the front entrance would go out of business long before the health inspectors arrived. However, if a business was forced by regulations to suddenly bear all the costs which had previously been borne by others, this would result in a sudden increase in costs which many firms would be unable to absorb. This would lead to a leap in unemployment, resulting in the closure of a significant percentage of firms, and possibly economic meltdown. It should therefore be appreciated that, at least in the short term, some externalities are inevitable.

A major problem associated with externalities is that a firm can deny any responsibility or liability because it is often:

- difficult to prove damage;
- difficult to identify the culprit;
- difficult to quantify the damage and therefore to put a monetary value on compensation.

Sandbach (1982) illustrates these points in a review of the Alkali Acts. In the modern world the notion of 'the polluter pays' is supposed to reduce pollution by forcing a business to internalize the externality. In other words, to turn the externality into a cost stimulates the firm to reduce or eliminate pollution at source. The case of the *Exxon Valdez* oil spill in Alaska (1989) resulted in what became known as the Valdez Principles. This meant that the polluter was obliged to pay for the damage caused, as measured by contingent valuation methods. Increasingly, governments and firms are turning to the principle of Integrated Pollution Control (IPC). This aims to incorporate comprehensive assessment and control of the environmental impacts of industrial activity. IPC has three dimensions, according to Blowers (1993, pp. 14–17):

1 The trans-media nature of environmental processes: pollution can affect air, land and water, so it is no use devising policies, say, just for water pollution without examining possible impacts on other media.
2 The trans-sectoral dimension: environmental pollution may arise in one area but have impacts in another, for example, changes in transport to curb pollution also have impacts on energy use, which will affect the energy sector. In the same way changes in one branch of a firm will have impacts on another.
3 The trans-boundary aspects of pollution: pollution crosses boundaries and is no respecter of jurisdictions. Therefore there needs to be adequate coordination between firms, local authorities and national governments.

A beach provides an example of the difficulties arising from externalities, that Hardin (1968) terms 'the Tragedy of the Commons'. When it is virtually empty, it has high value for its users since it is close to its natural state and any other users do not spoil the view. There is no litter or erosion. But as the beach becomes more widely known it attracts more visitors who bring with them litter, erode the dunes and paths, attract ice cream salesmen and noise levels and clutter spills over into the approaches which become congested with cars. The very qualities of solitude are being destroyed by increased use by people trying to get away from it all. Imposing some form of rationing, so limiting access to the beach, is probably the most practical resolution to this dilemma. The example could be followed of only so many tickets per day being issued, a system employed both by the French authorities managing the prehistoric painted caves of the Dordogne and by some National Parks in the USA. Other rationing solutions might be to impose increasingly higher parking fees or admission fees to deter visitors or to provide counter-attractions elsewhere.

Externalities can also exist spatially. A source of pollution can have spatial limits. Not everyone in a country is polluted equally. There may be an externality field (Figure 3.2), around a factory. The externality field could be positive as well as negative so that a plant which cleans water after use releases a benefit to the community at large, e.g. reed beds around sewage works soak up heavy metals and other pollutants and also provide an enhanced landscape and a wildlife habitat. But the positive field will have spatial limits, e.g. clean water discharged into a river from a factory which installed clean-up equipment would only benefit the next community downstream since the benefits would not extend any further.

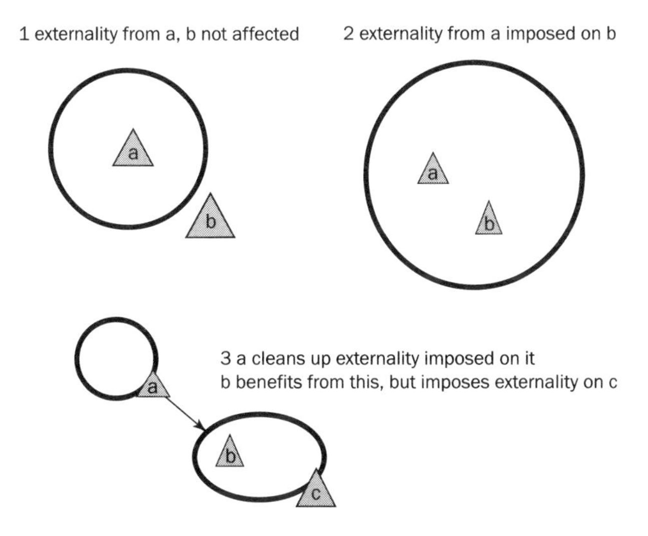

Figure 3.2 ***Examples of spatial externality fields***

Society, in general, and economists in particular, have long tried to resolve the problem of how to irradicate externalities. Take, for example, the case of the polluter who discharges effluent into a river and thereby reduces their internal private costs by externalizing the pollution. By this means profits are increased and a competitive advantage is gained over rivals, who choose or are unable to externalize their pollution costs. At the same time society is forced to either pay to clean up the pollution, or to take the consequences resulting from a degraded environment.

In the 1920s the English economist Pigou distinguished between the private costs of production (i.e. raw materials, labour costs, marketing costs) and the full social costs which reflect the full cost to society of activities and this includes the impacts of pollution. As the gap between private and social costs widens, so it becomes increasingly difficult for resources to be efficiently allocated, since the true cost of the resources becomes masked. Firms do not have to bear the full cost of production, choosing instead to externalize some element of the cost. This represents a failure of the market since the market price of the resource does not reflect its true cost.

Given that the manufacturer's cost should be reflected in the total costs – including the cost of pollution – a mechanism should exist for internalizing the social costs. There have been a number of ways in which society has sought to resolve this dilemma. Although a variety of solutions have been proposed, we have identified four central possibilities; two from the realm of the economist and two from that of the lawyer:

- A tax can be imposed on the polluter where the money raised through taxation would be set at the level exactly enough to pay for the damage caused. This is the use of the so-called Pigovian tax.
- A series of permits are issued allowing for set amounts of discharges and other forms of pollution to take place. These can be traded, thus allowing the market to determine who pollutes but within an overall envelope of total emissions. The market will therefore determine the least cost solution is adopted.
- The polluter can be made liable for any damage caused.
- The polluter can be prohibited from causing any (or at least excessive) pollution through the establishment and enforcement of emission regulations. A system of fines and legal action would be in place for those who exceeded the limits, although this is usually considered to be an economically inefficient solution to the problem.

All four present considerable practical problems if they are to be implemented. In the absence of complete information of the true social costs of the externality, setting the correct level of taxation has proved notoriously difficult. Set too low and pollution will continue to occur; set too high and the business is liable to go bankrupt. In either case the market would be said to have failed and the effect of the taxation would not achieve the desired outcome, i.e. the optimum level of pollution. Nevertheless taxation continues to be the basis behind the 'polluter pays principle'. This principle has acquired popularity as a method of pollution control in Europe and is being increasingly incorporated into UK legislation and policy, e.g. a Carbon Tax as a method of limiting the burning of hydrocarbons.

The other options are equally (if not more) difficult to implement. Pollution origins and destinations are both geographically uneven, but the concept of keeping global pollution at a given level is at the heart of the concept of emissions trading. Emissions trading is a way of giving some polluters a licence to pollute by buying quotas from other producers, who have reduced their emissions, perhaps through energy efficiency or fuel substitution. Quota trading allows the overall amount of pollution to be contained within some agreed maximum, but its distribution varies (Figure 3.3). For example, the UK has reduced its output of greenhouse gases largely as a result of the decline of the coal industry. It therefore has not used up its quota of pollution derived from a benchmark based on an earlier year's emissions. But another country might be unwilling or unable to reduce its emissions of greenhouse gases. It could therefore purchase and use the UK quotas to allow it to continue to pollute at existing or even higher levels. Even larger unused quotas are available from Russia and East European sources following the decline of manufacturing industry in the countries of the old Soviet Union and its satellites. The concept is not limited to the international stage, but works also within nations. The USA has a 'stock exchange' which trades 15 million tonnes of sulphur dioxide annually.

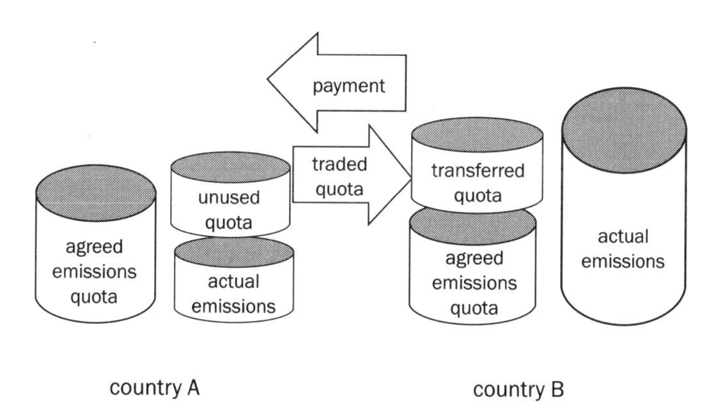

Figure 3.3 ***Emissions trading theory***

> On nineteenth May 1992 the Tennessee Valley Authority paid a Wisconsin power company for the right to belt several tons of sulphur dioxide into the atmosphere allowing the authority to legally exceed contamination limits set by law. Wisconsin cut its own pollution to offset Tennessee's.
>
> (Pallast, 1999, p. 4)

Permits to discharge pollution have become a tradable commodity and have resulted in a shift in the geographical distribution of emissions, while keeping the overall pollution total the same. It is an important feature of permit trading that those most efficient at reducing pollution will sell permits and the least efficient will find it necessary to buy permits.

If the polluter is made liable for any damage then the amount of damage caused has to be established in the courts. While popular in the USA, where there is a long history of using the courts to establish the level of such settlements, this approach has been shunned in both UK and Europe.

The final option we shall discuss is that of a regulatory limit. This approach has been favoured in UK legislation in the past. Although the approach adopted by the UK, which incorporates the use of the 'Best Practicable Means' has its advocates, it is not without problems (see Chapter 4).

It is perhaps worth pointing out the implementation of Pigovian pollution taxes does not result in zero pollution emissions, but rather seeks to establish the most efficient operation of the market by forcing polluters to internalize their costs and so achieve the optimal pollution level. It is rarely the case that no pollution is an efficient outcome. It is necessary to compare the benefits of lowering pollution with the costs of controlling it. It is likely that as pollution starts to rise, considerable benefit can still be achieved (profits for the firm, utility for society as a result of the benefits they gain from using the product). As the level of pollution continues to rise, the additional benefits which accrue will begin to decrease while the costs of control increase (production costs rise and it becomes increasingly difficult to find a market for the product. Labour costs rise and the capacity of the plant may be approached.) The optimum level of pollution is reached at the point when the net benefits (revenue gained from the product minus the private costs of production) exactly equal the combined cost of production and the costs of pollution. This is the level of production and pollution that Pigou was attempting to find. Beyond the optimum level of pollution, the additional damage caused outweighs the additional total good which is derived from the product, continuing to the point where there is no benefit from the final unit of production yet there are high pollution costs.

If the level of taxation is too low, then the producer will continue to pollute and the taxation raised will not be sufficient to make good the damage. If the tax is set too high, there is a danger that the firm may go out of business. More likely, however, is that they reduce their production to a lower level. This is to the disadvantage of the producer, who in reducing production levels reduces profitability. It also works against society at large since they do not have access to a product that could contribute positively to their well-being. From this we can conclude that Pigou's line of thought would solve the dilemma of internalizing externalities if only a mechanism existed for setting the optimum level of taxation.

There have been other attempts to resolve the dilemma looking at the issue in different ways. If we do not use the emotive word 'polluter' but rather 'community-based tradesman' to describe the producer, then a rather different story emerges. Imagine a local baker in a village. The baker provides considerable benefit to the local community. He provides local employment, wholesome bread and confectionery and along with other tradesmen provides a focus and sense of identity to the village. Customers may even come from surrounding villages to buy their bread, bringing further money into the local economy and leading to attendant multiplier effects. These are all positive elements of well-being or welfare to the community. However, by the very nature of his enterprise he makes a lot of noise early in the morning which disturbs those villagers living close to the bakery. This is an externality – there is an overall cost involved which the baker does not have to bear. Imagine one of the local residents was so disturbed by the noise he chose to move house and had to sell at a discount due to the noise. Now we have a measure of the cost of the externality. However, it would be not in the interests of the village community *in toto* to prohibit the baker from baking bread since the social benefits (are assumed) to outweigh the costs. Setting a taxation level or a subsidy to the local inhabitants (another possibility) would be fraught with practical difficulties. How can the dilemma be resolved?

Coase, the American economist and Nobel prizewinner for economics in 1990, saw the solution to this dilemma in terms of the property rights of the various parties, which are often ill defined. By property rights in this case we mean what the property owner is allowed to do – including emitting pollution if that is what they choose to do. It must be assumed that everything is owned by somebody and that all rights could be freely traded, then the market will behave in such a way that the optimum allocation of resources will take place and the level of externalities will ensure economic efficiency. It would not matter if either the baker or their neighbour were given the decision. If the baker were allowed to continue to make noise, then the neighbour would pay him to stop making a certain amount of noise. If the neighbour 'won', then the baker would pay so that he was allowed to make certain amount of noise. The end result would be the same – and the level of payment would ensure an economic efficient allocation of resources. The decision would merely determine who paid who – not the final result of whether noise was made or not. Although Coase's (1960) argument is compelling it suffers when such questions as 'who owns the air, the rivers or the seas?' are asked. Also Coase assumed the rights can be freely traded and there are no significant transaction costs – neither of which can be taken for granted in reality, especially if the property rights are held by a large and dispersed number of parties. While his thinking would resolve the dilemma surrounding our village baker, it would not provide a practical solution to many global environmental conflicts.

Subsidies and compensation

In many respects a subsidy is the opposite of an adverse externality. Both represent imperfections in the market. In an externality a cost is exported to a third party and in a subsidy a (financial) benefit is received from a third party. In neither case has the way in which the product or service is produced changed nor has any adjustment been made to the costs of production. Subsidies are imperfections in the market since they result in changes to the 'natural' use of resources. Usually they lead to more resources being used than would otherwise be the case. So subsidies to hill farmers or to wheat farmers result in greater production and greater uses of inputs than would be the case without subsidies. The distortion to the market can, unusually, have the opposite effect as with subsidies to farmers to take land out of production by setaside (see Chapter 6). Subsidies are usually provided because society has taken the view that social benefits, that the market does not recognize, will accrue and that the added benefit with the subsidies is greater than the value of the subsidy. Again, the upland hill farmer provides a good example. Without subsidies many hill farms would be abandoned and this would lead to a breakdown of rural life in such areas. This in turn would lead to a variety of detrimental social impacts, the costs of which are considered greater than the size of the subsidies – at least in theory, although the complexity of the situation makes detailed analysis subject to a number of interpretations.

Compensation can be looked upon as a particular form of subsidy between two parties. One has caused a disbenefit to the other and the compensation represents a settlement being reached, possibly using the judicial system, to the value (ideally) of the disbenefit caused. If negligence is shown, a company will be liable, if an employee or member of the public is killed or injured, or if damage to property has occurred. In some cases companies may have to pay compensation for *injurious affection*, that is, a nuisance such as pollution damage, which leads to the value of property being devalued. In the USA litigation can produce spectacularly large amounts of damages but compensation in other countries may be much less. The victims of Bhopal, for instance, had to wait a very long time for payment that usually amounted to hundreds rather than thousands of dollars compensation. The total amount of compensation demanded by the Indian Government was $3 billion, but the Union Carbide was in the end ordered to pay only $470 million (Elsom, 1992). Transboundary pollution compensation is a particularly difficult legal area as it involves different jurisdictions and may end up after many years being arbitrated by international courts. National governments may also be obliged to pay compensation to business if their activities are adversely affected. Farmers, for example, are compensated for loss of land and disturbance caused by road building and way leave for utilities such as gas pipelines (Blair 1981b). The airports fund insulation of homes from aircraft noise from Heathrow and Gatwick airports, an example of preventative compensation in business and another measure of the cost of noise. Most sound insulation in Britain, however, is the province of local authorities (Institute of Environmental Health Officers, 1991).

Transformations

It is in the nature of our subject matter that over a period of time the commercial worth of an item can change. What is an asset to a business, over time, with changes in product mix, technical innovation or changing demand, can become a liability. An entity, which at one time is a cost to a company and a drain on its finances, later can generate a revenue stream. Something that at one time is a cost to the company can be externalized so that a third party picks up the costs after a number of years. All of these possible transformations work in both directions.

It is perhaps worth remembering at this point the relationship between assets and costs. In the simplest of terms, assets are things which can be sold – they have a positive capital value. Liabilities are entities for which somebody has to pay to have removed – they have a negative capital value. Costs and revenue are about ongoing cash flows – they appear in the profit and loss account of a company. Costs results in money flowing out of the coffers of the firm. Revenue results in money flowing into the company. However, an asset can incur costs, but since it still has considerable capital value, a company will continue to incur costs in the expectation of a large capital return at a later stage. One such example is a newly planted forest. For the first decades of its existence it incurs costs and there is no revenue coming in. However, the forest represents a capital asset because of the value of the land and the anticipation that when the trees come to maturity they will be a realizable asset.

In the environmental field there are numerous examples of transformations which take place where the status of a resource changes. In terms of capital value through time, assets may move from being unrealized, to realized, to redundant, finally becoming a liability. Similarly, some liabilities may become assets. For example, we see the legend of Robin Hood and the sites he is purported to have frequented changing from a unrealized asset thirty years ago to a major tourist attraction for Nottinghamshire in the East Midlands, worth an estimated £67 million per year. Some assets when they are no longer worth anything can be seen as becoming redundant. Hence farms in the badlands of the American West in the 1920s were abandoned because they could no longer provide a revenue stream and could not be sold. In effect, they stopped being either assets or liabilities – they disappeared from the environmental assets register – to use a business analogy.

An example of an asset becoming a liability would be a nuclear power station at the end of its generating life. For most of its life while it is generating electricity the station would be regarded as a capital asset. However, as it reached the end of its life it becomes clear that it would be costly to dismantle, because of the radioactive contamination of its components, and it cannot be beneficially recycled. The plant's life cycle is over and nothing can be done to make it an asset. By 1989 estimated 'back-end costs' (i.e. decommissioning costs) of nuclear power stations in the UK were put at £13 billion. Total liabilities of nuclear power could exceed £20 billion since published costs do not include items such as monitoring radioactive wastes in the Irish Sea (which would be incurred with any decommissioning programme) or any future compensation or legal costs. The cost of decommissioning each Magnox power station has been put at £500 million and this was a major reason why nuclear power stations were not included in the electricity privatization of the late 1980s. No commercial buyer was likely to burden himself with a such a millstone (Rose, 1990, p. 138) without a commensurate source of revenue to meet such costs.

Liabilities sometimes have the ability to turn themselves back into assets. Gravel workings when they are operating are assets. At the end of their life they flood with deep water, representing a danger. They need to be drained, fenced and protected against fly tippers. However, it is easy to see this liability being turned into an asset. The lake could be sold or developed for a variety of sporting or leisure activities, including sailing, water sports or fishing. Good examples of such former gravel workings with new uses are to be found in the Colne and Lea Valleys on the edges of Greater London. Alternatively, a dry mineral pit completely worked out might seem to be a liability for the company that owns it. However, the company could charge a waste disposal company or authority to dump refuse in the pit. Once it is filled and consolidated, it could then be covered with aggregate, dressed with topsoil and then either sold or leased for agricultural use. Examples of this are the former chalk quarries near Tring in Hertfordshire and old sand pits near Leighton Buzzard, Bedfordshire. An even more enterprising use of derelict property is to turn it into an industrial archaeology or heritage museum as has been successfully achieved at Ironbridge, Morwellham and Beamish.

When examining the way in which ongoing costs can be turned into revenue streams there are numerous examples where the environmental potential has been realized. A good example of transforming a cost into a source of revenue is where a waste product is made into a new product. A sawmill produces vast quantities of sawdust, waste that presents a particular and costly disposal problem. Looking at the sawdust as a potential raw material changes the issue entirely. Sawdust can be used to make fibre boards and can be sold. With imagination and sometimes technological innovation, costs can be turned into a revenue stream. In Chapter 5 we pick up on this theme again.

The converse of the example above would be where the net revenue from an entity turns into a net cost. An example of this might be an upland farm where costs now exceed revenues from upland sheep. Upland farming for many years has been marginal in economic terms and almost all farms have only been kept afloat by either subsidies or revenue from diversified activities. In such circumstances the business cannot go on forever with costs exceeding revenue and where the land constitutes only a small capital asset. In such cases society and the business community can adopt one of two possible courses of action. They may choose to allow the abandonment of the land or they can institute a system of subsidies, which bring the total revenue up to a figure comparable to the total costs of the enterprise (including the living costs of those dependent on the farm). This is what has happened in the case of the upland hill farming. Those familiar with the example will be aware of the series of economic and social issues outside the scope of this discussion.

For the sake of completeness the reader should be aware of a further transformation that we have already examined – that of the movement between cost and externality. We have already discussed the way in which legislation and social pressures are forcing firms to internalize their externalities, hence changing an externality into a cost. However, there are examples of where the transformation is the other way around. Businesses are always looking to externalize their costs since it results in a reduction of the internal costs of the company. For example, in recent years there has been a tendency for computer companies to no longer provide manuals for their software and hardware products. This reduces company costs both for printing and distribution. The customer has to print out the manuals which requires them to buy the paper and to take on the responsibility for the use of this important renewable resource.

Further factors causing friction

One of the causes of conflict between business and the environment is the mismatch in temporal perspectives. Business typically has a short-term outlook while environmental processes are often very long-term indeed. Many significant environmental processes take place over very much longer time scales than business, but the problem is that business activities can interfere with the environmental processes. They can result in species extinction, destroying soil or fossil fuel reserves more rapidly than they can form or they can contribute to the acceleration of natural processes such as climatic change. An appreciation of the mismatch between the time scales is vital if business is to avoid creating new environmental problems. There are a number of factors that contribute to making the solution to the problems elusive:

- Identifying the future outcome of present action is difficult, e.g. no one foresaw the ozone layer destruction until many years after CFC's were first used. When first introduced, CFCs were regarded as a benign chemical since they are non-toxic.
- Definite benefits now are difficult to forego in exchange for indeterminate benefits sometime in the future. Responsibility towards generations unborn is a particularly difficult issue for business to view rationally. The present needs of shareholders and creditors weigh more heavily with a business than the possible environmental clear-up costs for society many years into the future. Since shareholders are in a position to look after their interests now, whereas future generations cannot do that, the near future gets more priority than the distant future.
- There is a danger in taking precipitate action. Past forecasts often turn out to be wrong and business has an understandable reluctance to take expensive action now, which may turn out to have been unnecessary in twenty years time. Many of the 'doom and gloom' predictions of the 1970s simply failed to occur. A BBC documentary of 1968 entitled *Due to Lack of Interest, Tomorrow has Been Cancelled* was a prime example of alarmist forecasting with several doomsters predicting complete environmental breakdown before the end of the twentieth century, which, reading this in the twenty-first century, you will realize has not happened.

Size matters

The size of business is another key variable that can influence the relationship between business and the environment. Large multinational firms tend to have explicit environmental policies because they operate under a number of legal codes in different countries and have experience where very strict environmental policies operate, such as the Scandinavian countries or The Netherlands. They are legally obliged to meet certain standards and to have transparent reporting mechanisms. Normally such a firm introduces the practice of environmental policies and statements across the whole corporation, even if it is not required to do so in every country in which it operates. The DETR (1998, para 49) points out that: 'In the UK most of the top 100 companies currently report on environmental matters but the second tier is not as vocal. Just over half of the next 250 companies provide information about their environmental performance.' Large corporations are also high profile organizations and are often in the news. It is a matter of public relations and image building to be seen to be environmentally responsible. Hence a company like Shell pays great attention to its environmental image (Box 3.3).

Box 3.3

Royal Dutch Shell environmental policy

No. 1 Pursue the goal of no harm to people.

No. 2 Protect the environment.

No. 3 Use materials and energy efficiently to provide products and services.

No. 4 Develop energy resources products and services consistent with these aims.

No. 5 Publicly report on performance.

No. 6 Play a leading role in promoting best practice in their industries.

No. 7 Manage health, safety and environment matters as any other critical business activity.

No. 8 Promote a culture in which all Shell employees share this commitment.

Royal Dutch Shell Group HSE policy

Every Shell company:

- has a systematic approach to HSE management designed to ensure compliance with the law and to achieve continuous performance improvements;
- sets targets for improvement and measures, appraises and reports performance;
- requires contractors to manage a HSE in line with this policy;
- requires joint ventures under its operational control to apply this policy and uses its influence to promote it in other ventures;
- includes HSE performance in the appraisal of all staff and rewards accordingly.

Source: Shell (1998b).

Small and medium-sized enterprises (SMEs), on the other hand, may be less inclined to publish formal environmental statements or indeed to actually have a formal environmental policy. It is difficult to say where a cut-off occurs but the archetypal small business, such as the corner shop or the family farm will be extremely unlikely to have any sort of written policy. These sorts of business will tend to be reactive to the environmental policies of governments and suppliers and to make environmental decisions as a result of imposed pressures, e.g. a farmer will respond to government incentives to adopt environmentally friendly farming practices if a grant or subsidy is involved. SMEs may individually account for small environmental impacts, but in aggregate they account for a great deal. Just as many other business practices have filtered down from the large to the small company, so one might expect that in future as environmental regulations become tighter so even SMEs will eventually come to have environmental reports. Not all small businesses should be seen as lagging behind their larger counterparts in the environmental field. Some small businesses are widely believed to be innovative and adaptive and many specifically new 'environmental businesses' can be categorized as SMEs (see Chapter 9).

Scale issues – spatial equity

Who is responsible for an externality? If there is an isolated factory in a rural area it may be easy to attribute blame, but what of a densely occupied urban area with numerous possible sources of pollution? Synergistic effects of pollution often render it hard to say who exactly is responsible for any particular damage. It is also the case that the social geography of an area determines the seriousness with which a pollution incident is taken. In middle-class suburbs, articulate and well-off people are more likely to obtain satisfaction from the judicial and political process than are people in poor inner-city areas. As was said in a celebrated nuisance case (Sturges v. Bridgeman) in 1879: 'whether anything is a nuisance or not is a question to be determined, not by an abstract consideration of the thing itself, but in reference to its circumstances; what would be a nuisance in Belgrave Square would not necessarily be a nuisance in Bermondsey' (quoted in Sandbach, 1980, pp. 49–50). In other words, it is much easier to pollute poor people than rich people.

On a larger scale, the controversy over acid rain falling on Scandinavia illustrates similar principles of the difficulties of attributing blame and acknowledging responsibility. Scandinavians suffering damage to their forests and lakes from acid rain blamed on emissions from UK coal-fired power stations. The British denied this for a long time and it was only reluctantly and grudgingly that it was acknowledged (Rose, 1990). The Norwegian Government refused to consider compensation issues as it believed that if polluters were effectively allowed to 'buy' the right to pollute, compensation would be a cheaper option for them compared to preventing the pollution at source (Elsom, 1992)

Measuring and valuing the environment

One of the chief problems that business encounters when confronted with the need to accommodate the environment in its calculations is that it is not always immediately obvious how to measure or value environmental costs, assets or externalities. Unless an entity could be bought or sold, business did not regard it as part of its concerns. The air was regarded as a free good. Anyone could use the air, it belongs to no one and therefore did not exist in the balance sheet. Equally, discharging noxious gases into the atmosphere was a free activity because who could complain about it? Not the 'owner' of the air because there was no owner. People, of course, did complain about smells and fumes but legally they had no redress unless they could prove that their property or business was in some way being damaged financially by the pollution. The common law notion of nuisance was often extremely difficult to prove. How could one identify the culprit if there were several factories emitting fumes? Which one of the many polluters was responsible for your trees dying or your cattle becoming sick and besides, who knows, they might have got sick anyway? If such tangible elements such as pollution damage were difficult to assess, the problem of intangibles such as the value of a view blighted by a road or a residential area exposed to aircraft noise was even more difficult to quantify. Although people objecting to a new airport might say that their peace and quiet was 'priceless', in practice they would be prepared to accept it if the compensation was adequate enough.

Environmental economists, such as Pearce *et al.* (1989), have done much to highlight the idea of valuation of the environment as a mainstream concept in various reports and books. Pearce and his colleagues have led the way towards incorporating the notion of valuing environmental goods. The problem is often put as measuring the immeasurable, or putting a market value on non-market commodities. A landscape is a classic case. What values do the view from Mount Snowdon or Mount Fuji actually have, compared, say, to mineral resources extracted from their slopes or grazing animals on their pastures? In a modern society the answer would be that we do value views above some tangible assets because organizations are prepared to pay considerable sums to protect the view from change. The National Trust was prepared to pay £4 million for the farm that included the summit of Snowdon, precisely because it was regarded as a national treasure. As a farm on rough, infertile, rain-sodden slopes, given the fact that all it could support were a few hundred head of sheep, at a time when farmers were making substantial losses on sheep, it should be hard to give the land away, let alone expect millions of pounds for it. But because it has value as a cultural icon, as a unique and beautiful landscape, it can command a high price. If the asking price seems excessive, consider that a Turner watercolour of the mountain would probably command even more. Environmental values are what society or individuals are prepared to pay. The difficulty is in agreeing what those values should be. In a competitive bid there is no real problem, if someone is prepared to pay millions for a mountain then so be it, that is what it is worth. However, what about putting a value on the cost of something no one wants such as pollution? How do we put a price on polluted water or radioactive waste? Is it simply the cost of clearing it up or the cost of potential health hazards? These questions become acute when legislators seek to impose pollution charges or fines on business to pay for pollution or to deter the polluter from polluting.

The growing demand by governments and society at large for business to account for environmental damage and the very real imposition of pollution charges, taxes and fines has led industry to develop green accounting. This means putting a price on the environment so that it can be incorporated into corporate accounts. The Society of Management Accountants of Canada (CMA), among other bodies, has produced a number of guidelines on green accounting, sustainable development and meeting various international environmental standards (CMA, 1998, online, see Box 3.4).

The accountants' problem in incorporating environmental factors revolves around giving a monetary value to things which do not always have a conventional place in the market. However, there has been a gradual process of increasing commodification of the environment with more aspects being given a price. A useful concept is contingent valuation which means essentially asking how much people would be prepared to pay for things which at present have no price, such as an

Box 3.4

Green accounting guidelines produced by the Society of Management Accountants, Canada

1 Tools and techniques of environmental accounting for business decisions – guide on how to measure, report and manage current and future environmental costs and opportunities. Costing analysis, investment analysis and performance evaluation (1996).

2 Implementing corporate environmental strategies – planning and implementing, improving and integrating environmental considerations into business decisions (1995).

3 Accounting for sustainable development – a business perspective (undated).

4 Accounting for the environment – identifying issues regarding environmental concerns and the role of management accountants (1992).

Source: CMA (1998).

endangered species. If it costs a certain amount to safeguard dolphins in a particular sea area, how much would people be prepared to pay to go dolphin watching in order to provide funds to achieve that aim or how much tax would they be prepared to see go to that objective? Despite the apparent attractiveness of deriving values direct from the public, Tietenberg (2000, pp. 39–40) points out that contingent valuation needs to be used with caution and has a number of potential pitfalls. First, there is 'strategic bias' which means that those with strong motivations for seeing high values emerge will deliberately overstate their responses to influence the result. Second, 'information bias' where respondents have little or no idea of the value of an environmental attribute and guess wildly. 'Starting point bias' is the third problem which occurs when the respondent is given set choices for valuation. The values ticked may depend largely on where the categories start – a survey that has steps of £10 will elicit a different result from one that has £50 steps. The final criticism is termed 'hypothetical bias'. Here the person being surveyed knows that the questions are hypothetical and are unlikely to result in demands for actual payments, so there is nothing to be lost by being 'generous' and bask in the glow of approval for being public-spirited. What people say in surveys or focus groups may not be what they would actually do in practice. However, it does provide some guidelines as what people think an environmental item is worth relative to other things, provided one is cautious about survey design.

Corporations are paying more attention to corporate environmental reports (CERs) because there is an increasing demand from legislators and stakeholders for information and accountability. A CER is a company's self-evaluation of its environmental performance and is the environmental version of the annual shareholders' report. Some companies include environmental factors in their annual reports, but it is a sign of the increasingly important profile of the environment that it is seen as necessary to have a separate CER. As more companies present CERs, so it becomes necessary to make them more inclusive and rigorous, since they will be subject to wide scrutiny and comparison with competitors' CERs. To avoid charges of 'green wash', a CER must be as thorough and transparent as an audited annual report.

Deloitte Touche Tohmatsu, the global accounting, auditing and management consultancy firm, has constructed a CER score card to evaluate the performance of CERs. It has based this on a manual of good practice devised for use in Scandinavia and it has now been tested on some 2,000 CERs and environmental sections of annual reports worldwide. Deloitte Touche Tohmatsu's (DTT) CER score card is therefore intended to provide both a way of evaluating published reports and comparing different companies, in addition to being a guide to best practice. The score card has forty sections under eight headings and shows what could be expected in the ideal CER (Box 3.5).

DTT point out that no company it has evaluated has got anywhere near the maximum score. In addition, the CER does not evaluate environmental performance, but rather its ability to report its environmental activities. Since CERs must be subject to verification, the attempt at producing such a comprehensive report must force the corporate mind to concentrate on issues which it has perhaps not confronted before. This may lead to real environmental improvements. Just as EIA statements often lead to more rigorously thought-out development proposals with a lower environmental impact than would otherwise be the case, so CERs may well be a force for environmental improvement in business. It does not necessarily mean that this will happen, but avoidance of public embarrassment is a strong motive. The transparency of the CER score card is a highly positive feature. More and more companies are making a CER, as distinct from an environmental

Box 3.5

Deloitte Touche Tohmatsu's CER score card system

Reporting categories

A Corporate profile
1 Corporate context – about the company
2 Management commitment to the environment
3 Consideration of significant environmental aspects
4 Environment policy and commitment

B Report design
5 Scope of the report, including its limitations
6 Rationale behind choice of environmental performance indicators (EPIs)
7 Reporting and accounting policy
8 Description of relatedness and pertinence of environment to management and finance
9 Coverage – extent of activities, issues and achievements

C Environmental impact/data
10 Inputs – consumption of materials, energy, recycling, efficiency measures
11 Emissions – list of emissions and efforts to limit them
12 Waste/residual products – types and methods of disposal
13 Packaging – types, amounts, efforts to minimize
14 Transportation – environmental impacts and efforts to decrease
15 Product stewardship – environmental impacts of consumption and disposal, measures taken
16 Land contamination – reduction of environmental burden and remediation
17 Environmental effects – of operations, products, services – local, regional, global
18 Other factors – e.g. work environment, community involvement

D Environmental management
19 Environmental goals and targets
20 Environmental management systems
21 Integration into business processes
22 Compliance – with environmental regulations – details of how it is done
23 Contingency planning and risk management
24 Research and development – how is environment integrated into R&D
25 Life-cycle design – how is environment integrated into life-cycle design

26 Environmental Impact Analysis – how is EIA used in company projects, investments, etc.

E Finance/eco-efficiency
27 Environmental costs/investments
28 Environmental liability – e.g. clean-up arrangements
29 Government economic penalties/incentives
30 Future costs/investments
31 Business opportunities and risks – benefits of environment/hazards
32 Eco-efficiency measures – correlation between environmental and financial performance

F Stakeholder relations
33 Employees – degree of involvement/education/suggestions/feedback
34 Customers and consumers – environmental needs and preferences
35 Contractors and suppliers – environmental needs and requirements
36 Regulatory bodies – involvement with and relationships
37 Voluntary initiatives – involvement with environmental NGOs

G Communications
38 Layout and appearance of report
39 Communication and feedback mechanism

H Third party statement
40 Third party statement – external auditors' or consultants' verification

For details of scoring system and full explanation of types see DTT's website at: http://www.teri.tohmatsu.co.jp/services/scorecard_E.html

appendix to the annual report. In some countries this is a legal requirement and may well be expected to spread to others.

Cost benefit analysis is probably the most widely used economic technique for evaluating projects. It is based on the principle that if a project is to proceed the total benefits should outweigh the total costs. In the event of more than one option fulfilling this criterion, the preferred project is the one with the greatest difference of benefits over costs. In any scheme there are likely to be both winners and losers. So, for example, with a new airport most people will be winners in that they have a new facility and the airport will bring business to many people. However, those living in the flight path and those at rival airports who will lose business as a result of the new airport will be losers. There has to be some way in which the sum of all the benefits and costs can be brought together to determine if the scheme should be proceed. Pareto, the Italian economist working around 1900, looked at under what circumstances society would be better off and devised the set of rules for such a condition. This has provided the foundation for what we now know as welfare economics.

Cost benefit analysis (CBA) (or benefit cost analysis, as it is known in the American literature) has now become a highly technical subject that essentially requires that all costs and benefits, both tangible and intangible, are expressed in monetary units. It is the measurement of the intangibles that presents some of the most difficult problems. There are numerous texts on the subject of cost benefit analysis which approach the subject from a range of assumed knowledge from the elementary level to the highly technical and mathematical. At one end of the spectrum are texts such as Turner, Pearce and Bateman (1994) which is aimed at the general academic reader to that aimed at economics undergraduates (Pearce and Turner, 1990) to those of the graduate and professional market (Hanley and Spash, 1993).

While CBA is well developed in a number of areas, the environment presents a particular set of problems. It is not easy to place a value on an idyllic landscape or a healthy climate. A number of highly sophisticated techniques have been developed to overcome these problems. One such is the willingness to pay where it is estimated how much an individual wishes to pay to use or retain a feature. Another technique is that of the subsidy in which a party is offered a subsidy to forego a good. A further issue which the techniques must address is how the value of something today can be offset against future costs. This final technique requires some measure of any value today to be discounted into the future – an area of much detailed work, but which is neither necessary nor appropriate for us to go into in this book.

Summary

The reader should now be aware of the distinction between environmental costs, assets and externalities in business. All businesses try to minimize their costs, maximize the value of their assets and make a profit in order to stay in business and grow. It is possible that a business retains other goals. For example, the business may support charitable aims or may be allowed to make a loss because of some wider community or national interest. It is true that some businesses operate at a loss and are subsidized by the government for social reasons (e.g. start-up businesses in traditional coal mining areas) or for environmental reasons (e.g. payments to farmers in environmentally sensitive areas). However, even in these special cases, the aim of the business is still to minimize costs and to increase its assets.

The difficulties in identifying and measuring the assets, costs and externalities should be apparent, as well as the methods used in trying to evaluate them. The ways in which each of the elements can be transformed into each other should also be realized. Transforming externalities into internalized costs is the result of complying with environmental regulation or yielding to either peer or consumer pressure. Businesses have a huge repertoire of methods to reduce costs. These include gaining benefits from economies of scale, agglomeration, technical innovation, managerial innovation, process substitution, material substitution, product diversification, mergers, acquisitions, assets sales, relocation to cheaper sites at home or abroad, reduction in labour, energy efficiency and conservation, recycling and automation. A business faced with the need to reduce costs or internalize an externality is frequently ingenious. Provided the business is sufficiently motivated to act, it will usually find a way to solve the problem.

The key difference between time scales in business and the natural environment helps to explain some of the problems business has in addressing environmental issues – ten years is a long time for a business, but a mere eye blink for the natural environment. Who pays for the environment is a further vexed question and the problem of resolving public and private contributions makes finding solutions difficult. Putting a price on the environment whether through taxes, fines, quotas, compensation or commodifying pollution is not easy, but increasingly governments and business are trying to do just that.

Further reading

There are a number of useful chapters in M. Jacobs (ed.) *Greening the Millennium? The New Politics of the Environment* (Oxford: Blackwell, 1997), which put environment and business into a political context. P. Howes, J. Skea and B. Whelan, *Clean and Competitive? Motivating Environmental Performance in Industry* (London: Earthscan, 1997) takes an eco-efficiency approach to show how businesses are motivated above all to save costs. It also reveals just how poorly many businesses are at accurately judging environmental costs. Spreading good practice and adopting standard methods are important in measuring and valuing costs, assets and externalities. A useful method which is widely used around the world is the Deloitte Touche Tohmatsu (1998) *Corporate Environmental Report score card: A Benchmarking Tool for Continual Improvement*, online. Available http://www.teri-tohmatsu.jp/services/scorecard_E.html. Accountants are increasingly coming to terms with environmental accounting and auditing as shown by D. Owen (ed.), *Green Reporting: Accountancy and the Challenge of the Nineties* (London: Chapman Hall, 1991).

D. Pearce, A. Markandya and E. Barbier, *Blueprint for a Green Economy* (London: Earthscan, 1989) is worth looking at to consider the wider aspects of a green economy. S. Eden, *Environmental Issues and Business: Implications of a Changing Agenda* (Chichester: Wiley, 1996) focuses on the corporate scale and examines many examples of how businesses actually approach environmental issues.

Discussion points

1 What are the main theoretical and practical difficulties in establishing accurate values in cost/benefit analysis?

2 How would you use contingent valuation to establish public attitudes towards a proposal for a theme park on public recreational land?

Exercises

1 Evaluating Corporate Environmental Reports

1 Access the Corporate Environmental Reporting website (http://www.cei.sund.ac.uk.envrep/corprepS.htm) and select any report from whichever company interests you. Evaluate it on the basis of the CER score card shown in Box 3.5.

2 Go to any firm's website and see if they have an environmental report. Evaluate it as above.

3 Depending on the amount of time you have, or if working on your own or in a group, a number of exercises could be undertaken relating to CERs, for example:

(a) comparative evaluation of several companies in one sector, e.g., oil;
(b) comparison of different sectors, e.g. forestry, retailing, microelectronics;
(c) comparisons of different countries in the same industry, e.g. compare Australian, UK, American and Scandinavian firms.

The only limits are your imagination and time. These exercises could form the basis of a seminar, essay, mini-project or an undergraduate dissertation.

2 Costs, assets and externalities

Draw up a matrix to identify costs, assets and externalities for each of the following: coal mining, chemicals, railways, airlines, hotels, telecoms.

4 Environmental business necessities: the pressures which cannot be ignored

- The relationship between business and environmental goals
- Real, anticipated and perceived pressures
- Pressures on business from the past, present and future
- Physical pressures
- Why direct environmental pressures are often overlooked by business
- Changing environmental conditions
- Political pressures
- Environmental legislation
- Legislative enforcement and interpretation
- Economic pressures
- Socio-cultural pressures
- Technological pressures
- Environmental impact assessment (EIA)

> *Environmental concern has created clear winners and losers in industry. While some corporate environmental policies are compatible with commercial self-interest, others have been driven by the fear of environmental liability and failure, with levels of environmental investments outweighing any immediate gain.*
>
> (Howes *et al.*, 1997)

While seeking to fulfil their goals, businesses are subjected to a series of influences and pressures. Some of the influences are external to the company; these include market trends, government legislation and the action of their competitors. Some are internal, such as the company's investment (or lack of it) in R&D (Research and Development), their allocation of funds and the success of their marketing strategy. In addition to forces directing the firm towards its business goals are a series of pressures which force the company in other directions. One of these is towards certain environmental goals. This chapter examines the ways in which external pressures with an environmental dimension influence the behaviour of the company and identify how these pressures come to bear as a result of the various influences upon the company.

The interaction between business and the environment is a complex one. A company can be regarded as being subjected to a number of external influences. These are usually described in the business literature as being economic, social, political or technological. Since these influences are external, the company has little control over them. However, it can adopt a variety of strategies in response to these influences. Under some circumstances companies can turn these factors to their advantage. There is a range of such opportunities and the ways in which they can be exploited are explored in Chapter 5.

Although companies can do little about the fact that they are subject to external influences, they do have some flexibility in the way in which they can respond. Among the more obvious responses they can make are:

- plan and incorporate into their structures and processes a systematic response to the pressures;
- capitulate and cave into the pressures taking whatever businesses costs and disbenefits are incurred with such a course of action; the action of Shell in response to environmental criticism to its disposal of the Brent Spar platform (see Box 3.1, p. 68) could be seen as falling within this category;
- stop the pressure being applied in the first place by pre-emption and appropriate lobbying (this makes the assumption that the company has the necessary power to be effective in lobbying), for example, the National Coal Board as early as the 1970s incorporated into its planning Environmental Impact Statements (see later this chapter) to head off opposition to various proposals long before they were required by legislation;
- take out insurance in one form or another, including self-insurance, diversification of product range, etc. to reduce the impact of the pressure on particular activities. The investment made by the big oil companies in alternative technologies such as solar power can be seen as a form of diversification against the day when economic or perhaps political pressure prevents the continued reliance on oil as the main income stream of these companies.
- be pro-active and turn the applied pressure to the company's advantage, a strategy which is explored in depth in the next chapter.

The relationship between business and environmental goals

The prime aim of any business is to achieve its business goals and the activity of the business is directed to these goals (see Figure 4.1). They will achieve these goals essentially by their own driving force, but will also be influenced by external factors which the successful firm will be able to either turn to their advantage or to assimilate into their business processes. In the business literature these external influences are often examined in terms of a PEST analysis (or STEP analysis if a more positive sounding word is preferred). In either case the initials stand for political, economic, sociological (or more correctly socio-cultural) and technological. Not only do these external factors influence the company in how it goes about achieving its business goals but they also influence how the company achieves other objectives including those with an environmental dimension, such as:

- reductions in environmental degradation;
- greater sustainability;
- an appearance of being 'green'.

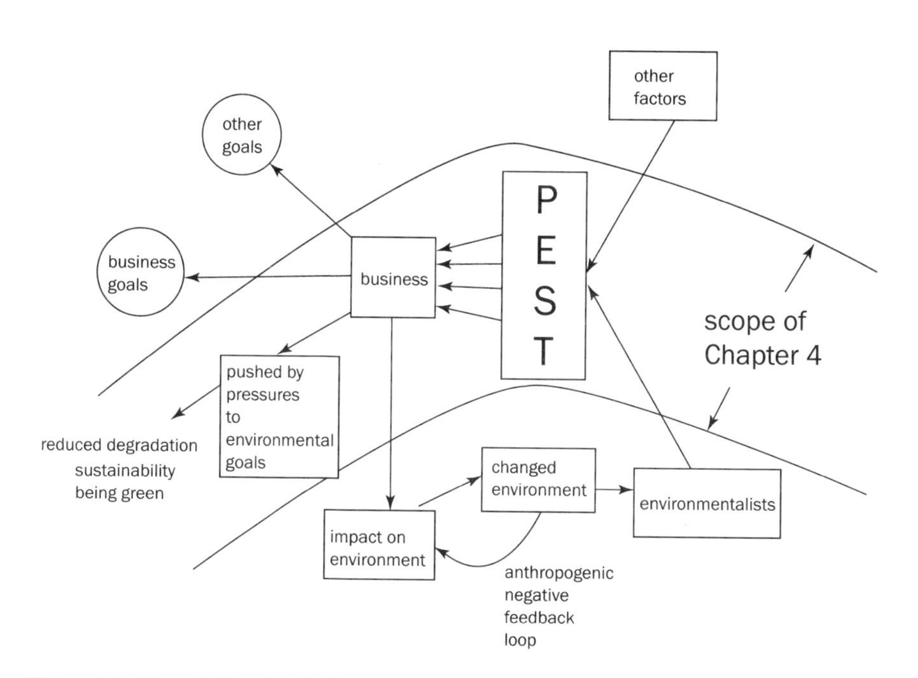

Figure 4.1 ***The relationship between external pressures on business and behaviour***

At the same time as the business is being pushed towards particular environmental goals it will have an impact on the environment through its operations. It will cause pollution, use energy, use up natural resources and these will have an impact, which is illustrated in the lower half of Figure 4.1. The cumulative effects of these changes result in the feedback loop of environmental degradation. This directly impacts on business in ways explored later in the chapter. The fact that business results in environmental degradation is seized upon by environmentalists. They then use this as evidence to put pressure on business to modify its activities and hence complete the cycle illustrated in Figure 4.1.

Businesses are subject to pressures that are generated externally to the company. These are the result of the four PEST influences, to which we have introduced a fifth, that of the state of the physical world. It is our contention that companies are subject to a series of physical environmental influences that are part of the external environment and to which they must respond. These external influences originate from a number of different sources. In an advanced capitalist market economy it is business itself that generates many of these influences, as shown on the right of Figure 4.2. Many of the influences derive from business directly. Hence economic pressures come from the competitiveness generated by business rivals, or from legislation designed to maintain the order and stability of the world economic and commercial order. Even technologically based pressures result from the entrepreneurial skill of

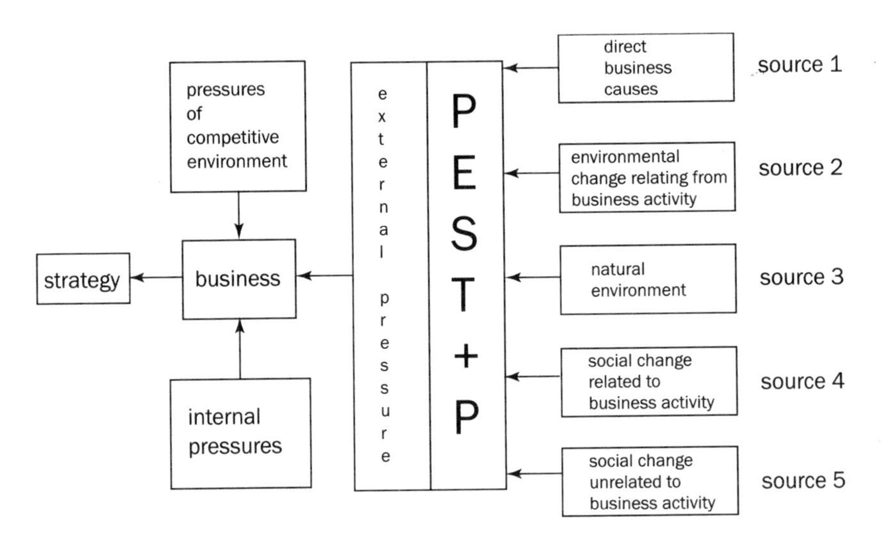

Figure 4.2 ***The origins of the external influences upon business***

manufacturers making available new and better machines for the processing of materials or from technological advantage which is gained by more accurate and timely information provided by more advanced computer technology. So if one company has adopted new technology to become more competitive, then there is pressure on rivals to adopt similar technology.

Second, a business may be subject to environmental change caused directly by business activity. Such an example would be where one company, perhaps a fish farm, was polluting a water course which destroyed the fresh water supply for a second business, perhaps a water cress farm, further downstream. It is more usual, however, that business activity in general results in environmental degradation, e.g. atmospheric pollution, increase of lead in the atmosphere. Society, then, exerts a combination of social, economic and political pressures on business to redress the degradation. Here the pressure is not a direct physical pressure although the source of that pressure can be traced back to a physical change.

While business, or at least society, has generated most environmental influences on business, some are the result of natural environmental change, and these are shown as source 3 in Figure 4.2. The direct influences of the physical world have already been considered in Chapter 1. Very often it is difficult to separate human from natural causation. To illustrate this point, consider the pressures that may be put on a manufacturing plant as a result of rising sea level. Some authorities will attribute the sea level change to increasing industrial activity resulting in additional greenhouse gases, which leads to a rise in atmospheric temperatures that in turn melts the ice caps and so causes sea levels to rise. This represents a sequence of changes that can be attributed directly to business activity. On the other hand, it might be argued that such changes can be attributed to the natural cyclical behaviour of the Earth's systems and would have happened even without the intervention of man, as has occurred throughout geological time. In reality, causation is probably a combination of both, although to the company this is of little consequence. Their prime concern is that they are subject to an environmental pressure to which they must respond in some way.

Not all pressures on an individual business, which ultimately result from the actions of other businesses, can be categorized as either business or environmental in origin. These are included for the sake of completeness as sources 4 and 5. Many are social in nature. One such example is that of child labour. In the nineteenth century many factories employed child labour. Social reformers who lobbied employers directly seized upon the social implications of this. They were instrumental in introducing legislation into Parliament and raising public awareness of the inhumanity of such practices. The combination of these actions put pressure on employers to change the conditions of employment for the very young. Reforms such as these were often bitterly resisted by businesses at the time, but gradually became accepted as normal elements of the cost structure. In a similar way, initial proposals for environmental regulation are frequently opposed by business, but are eventually accepted into the agreed framework of operations and costs once they have become law. Many social pressures that come to bear on business have an environmental dimension. Some of these can be traced back to business activity (source 4), e.g. pressures on supermarkets to stock organic food lines, while others are independent of business activity (source 5), e.g. social pressures relating to the right to roam across open countryside.

Real, anticipated and perceived pressures

Companies are subject to a range of different pressures in the external environment to which they must respond. Some of these are tangible such as the cost of raw materials, current legislation or the terms of international trade. Other are less tangible but nevertheless important. Sometimes the pressure comes from anticipating changes that have not yet happened. For example, one of the concerns of businesses in the run up to the millennium was to ensure that their computer systems would continue to function correctly after 1 January 2000. This was a pressure created by an anticipated event. Sometimes the pressures may be more nebulous. For example, if it were thought that there would be a move away from car-borne transport of individuals in the future, retail companies might begin to respond to this perceived social pressure by locating new outlets near public transport terminals rather than near motorway interchanges.

Whether a company decides to commit resources to evaluating an influence is a good test as to whether that external influence is regarded as something that needs to be heeded. The company is in effect saying, we cannot afford to ignore this factor in the external environment over which we have no control. This is a measure of how important the physical environment is to a company. Hence, companies carry out competitor analysis, market research into social trends, appliance ownership and changing lifestyles and they invest in services which give them expert advice on likely meteorological conditions in the future – a clear environmental influence. Different businesses are subject to different external pressures. A hairdresser in the archetypal UK High Street will not monitor world commodity and exchange rates – but a multinational oil company will. A French farmer will not monitor earthquake activity but a Californian water utility close to the San Andreas fault will. Different companies have different pressures but all have to be prepared to commit resources to measuring the external pressures to which they are subject.

Pressures on business from the past, present and future

Most of the thinking of this chapter, so far, has been directed at the present. Implicitly we have been thinking about pressures that are being applied as a result of present-day circumstances. There is, however, a historical dimension to all of this. Companies are subject to pressure from decisions that have been made in the past. Such decisions were often made at a time when political, economic, social and technological conditions were very different and invariably when attitudes to the environment were also different. For example, power stations are now under pressure to retrospectively fit scrubbers to their emission systems to reduce sulphur emissions into the atmosphere. The technical design decisions were made at the time the power stations were built although the social pressure is being applied today. Companies maintain they are unable, for either economic or technical reasons, to comply with demands for such changes; nor are they able to start again with a clean sheet and to go back and remake decisions made in the past.

Just as a company comes under external pressure from the results of business activity and decisions made in the past, they also come under pressures to justify what they are going to do in the future. There are a whole series of external pressures that force companies to evaluate the environmental, as well as the business, consequences of what they are planning to do in the future. This requires companies to carry out environment impact assessment (EIA) and to consider the total costs and benefits that any proposed action is likely to have. We will briefly consider the way in which the pressures to predict the environmental impacts of business activity in the future are worked through towards the end of this chapter.

The way in which an individual company is forced to respond to external pressures, which ultimately came about from the company's and other businesses' activities, have been described. When looked out through time the cycle can be viewed as a spiral, as shown in Figure 4.3. Using the example of air pollution and its control at point (a) (in the early 1850s) companies were emitting noxious pollutants out of their chimneys. This resulted in air pollution (b) and a rise in public concern (c). The concern was such that legislation in the form of the Alkali Acts was passed and a political pressure was exerted on the companies not to emit the noxious substances. So conditions started to improve and by (d) conditions were better. Although they no longer emitted the noxious wastes, they continued to emit smoke as a result of coal burning. This resulted in soot emissions and air pollution (e) which manifested itself in the smogs of the mid-twentieth century (f). Public concern and a link between smog and illness (g) led to the passing of the Clean Air Act of the 1950s. We can see this as a rerun of what had happened before with political pressure being put on companies to reduce their smoke emissions. This had taken place by (h). By the end of the twentieth century we were into a third loop of the spiral with public concern being raised (i) about the harmful effects of the release into the atmosphere of sulphur compounds resulting in acid rain which is having a detrimental effect on our cities (j) with already pressure beginning to be exerted to change industrial practices and new laws coming into force (k).

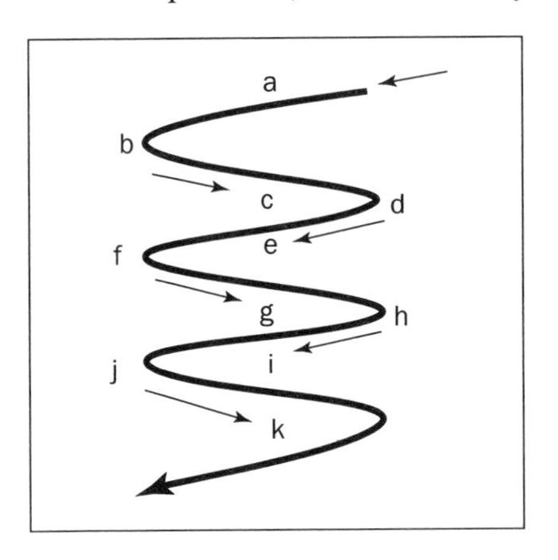

Figure 4.3 ***The business response cycle***

Physical pressures

Although little attention is given to the way in which the environment influences businesses in most of the literature, there are examples where they are of considerable importance. This section will identify a number of ways in which the natural environment can influence a business, which result from one of three broad causes:

1 the absence of physical resources (raw materials or natural elements to produce raw materials); this usually comes about due to the raw materials having been used up;
2 presence or absence of land with the required physical attributes;
3 changes in the prevailing environmental conditions. These changes may be systematic and predictable or sudden and unpredictable.

Most influences are predominantly long term, such as those relating to changing climate, sea level or soil structure changes. Sometimes they are the result of infrequent but catastrophic events such as floods or earthquakes. A distinction should be drawn between a real change that exerts an influence on business and a perceived change for which society legislates and with which a business is required to comply. For example, a distinction should be drawn between the following two circumstances. In the first, rising sea levels will directly force a coastal business to move or carry out some flood protection work. In the second, a government demand for a company to reduce its emissions of carbon dioxide and so contribute to international targets of greenhouse gas reduction intended to mitigate global warming, will indirectly help to prevent sea levels from rising. This by implication is done in the anticipation it will prevent flooding.

The natural world is constantly evolving and while a state of dynamic equilibrium exists, nature is not static. The natural adaptation or evolution of insects to pesticides, for example, means that some insects have become resistant to certain pesticides. This requires new varieties of pesticide to be developed or other measures taken to overcome or circumvent the problem. Hydrological variations in precipitation amounts and distribution coupled with naturally varying groundwater and run-off levels have resulted in changes in agricultural practices. In addition, manufacturing businesses have been denied reliable supplies of water. Rising sea levels on some Pacific atolls raises fundamental questions about the viability of island communities to continue and the retreating coastline of eastern England is having a real impact on the farmland and those businesses adjacent to the coast. As populations grow, the physical limit of some migratory agricultural practices in the drier parts of Africa puts an absolute physical limit on the available physical resources. The most efficient agricultural practices in such areas rely on shifting cultivation or nomadic grazing, which have physical limits on the numbers of people and animals that can be supported. When this total is exceeded, there follows a progressive breakdown of the limited fertility of the land and the prevailing agricultural system.

Physical limits can be exceeded in other ways. Hence the exhaustion of minerals may result in businesses needing to seek exotic sources or materials substitution if they are to continue in business. The disappearance of physical resources is not restricted to minerals. Overfishing has resulted in the demise of the North Sea herring industry. Agricultural lands in what is now the badlands of the plains of the United States have now been abandoned. The tertiary sector is not immune from such depletion either. For example, the decline of big game from the plains of parts of Africa has brought a premature end to the potentially lucrative tourist industry. The above examples in the main are the result of changes in the natural environment. Most are characterized by pressures on business resulting from the rate of change being greater than the business can cope with or where sudden, unexpected and catastrophic events occur with greater regularity than the businesses can manage. Most of the environmental changes may not be totally 'natural'. In most, human activity has either brought on directly, or has been a contributory cause of, the change. The burning of fossil fuels has undoubtedly contributed to the plight of Polynesia where the sea is slowly inundating atolls, although whether this is totally due to man is not certain.

Why direct environmental pressures are often overlooked by business

For many businesses the physical environment may not seem that important. Most basic business activity is not dependent on the workings of the physical environment. A second factor is that modern society has isolated us from many of the environmental excesses and unexpected changes. Heating, ventilation, modern buildings, transport links are designed to distance us from the ravages of nature. Third, most physical change, as opposed to catastrophes such as earthquakes or tornadoes, tends to be relatively slow. For example, even if global warming has dramatic effects, these are going to be a slow – imperceptibly slow within our lifetimes. Such changes are not going to influence business with their planning horizons measured in months or years rather than decades or centuries.

Since primary industries are closer to the physical environment, the direct pressures on them resulting from the environment are likely to be more obvious and this will be considered in more detail below. For the manufacturing sector of industry the natural environment rarely exerts a direct pressure. A shortage of natural raw materials might be considered a direct influence. This is only so in the unusual event of the absence of an essential raw material. Where raw materials begin to run out or be in short supply, it is not a physical pressure on the enterprise that is noticed. What happens is that if demand outstrips supply the price rises. The materials are still available (and so there is no absolute physical pressure) but are more expensive (so an economic pressure is exerted). This, after all, is what economics is – the allocation of scarce resources. In such a case the firm can adopt a series of tactics. It may invest in a different technology in its manufacturing process (which exerts a technical pressure) or it may look to secure supplies from more dubious sources i.e. from politically or environmentally sensitive areas, in which case it may subject itself to social pressure.

For many primary industries, however, the substitution of one source of supply by another, albeit at a higher price, may not be possible. For the fisherman when catches decline, the presence of raw materials is a real pressure as he is totally dependent on these supplies. Sometimes there is a perceived shortage in which case the shortage is created by another of the PEST pressures rather than an absolute shortage of the resource. For example, the imposition of fishing quotas or the restriction in the mesh size of nets are political pressures on fishermen. Where the public turns against tuna as a food source, on the grounds that the fishing techniques result in the death of dolphins, this should be regarded as a social pressure. However, the environment will exert a direct pressure if the number of fish the fisherman is able to catch, having allowed for any substitution of species or other remedies open to him, still are not sufficient for him to sustain a business activity.

There are numerous cases quoted in the literature (Goudie, 2000; Owen and Unwin, 1997; Simmons, 1996) where the exhaustion of physical resources has led to businesses changing the way in which they operate. When almost any aspect of the physical environment has changed, there will be some businesses that can no longer ply their trade. Among the elements of the physical environment which can change are:

- geomorphic elements – slope failures, landslides;
- biosphere, including reefs, rain forest, fish stocks;

- hydrological – river and groundwater events, floods and drought;
- atmospheric and climatic change;
- edaphic conditions.

Each may be associated with the disruption of some economic activity.

In addition to the absence of physical resources, there is a second group of direct influences on business that may be overlooked. These relate to the availability of land with particular physical attributes. Sometimes degradation of the land has taken place to the point where it is no longer physically capable of being used, as it was originally, for a commercial purpose. The most obvious example here is where soil erosion has made land unusable for commercial farming or where mineral resources have been worked out, leaving a company with no source of raw material for a manufacturing activity. Just as real, however, are cases where either historic buildings or archaeological artefacts are lost from a site, or where despoliation means that the tourist potential of a site is lost. In all these cases it is the physical attributes of the sites that give the location its value for businesses activities. Stonehenge without the stones would have important commercial as well as heritage implications for English Heritage or the building of a major industrial complex on the North Cornish coast would have important implications for the tourist potential there.

Changing environmental conditions

Two categories of environmental change have important implications for business. First, there are systematic changes, which tend to be slow in taking place but which necessitate structural changes to business. Global warming, rising sea level changes and the shift in climatic belts fall into this category. All have important implications for business and require them to monitor such changes and to take action accordingly.

The second category of changing environmental conditions is the unexpected and often unsystematic onset of often catastrophic conditions. Catastrophes, by their very nature, are unexpected and it is the inability to predict and plan for such events which amplifies their effects. Storms, floods, hurricanes, droughts, intense cold or hot spells, volcanic eruptions, earthquakes, tsunamis, even plagues of locusts or forest fires fall into this category. In most cases, where such events have been anticipated and resources are available to put plans into operation, their impacts can be mitigated. The different impacts of hurricanes in the

Caribbean and cyclones in the Bay of Bengal highlight the contrasts. When hurricanes reach the US coastline, as they do each late summer, they result in world news, considerable damage to property but comparatively little loss of life. In contrast, when tropical cyclones hit the east India and Bangladesh coasts, deaths in thousands or tens of thousands are not unknown. In 1999 two severe earthquakes struck Turkey. In the first, rescue plans were uncoordinated and the Turkish Government came under considerable criticism, both from within and outside the country. By the time of the second earthquake much had been learned, and while there was still considerable damage and immediate loss of life, it has been acknowledged the longer-term impacts were less due to the coordinated rescue and reconstruction plans. It is not within the remit of this text to describe or even catalogue such events in any detail. However, the reader must be aware that they can have serious impacts on business. It is usual for public attention to be focused on the human suffering immediately after such disasters. However, they result in considerable commercial disruption, which lasts long after the media spotlight has disappeared.

Any business that pays money or puts resources (of any kind, equipment, human effort, research, etc.) to find out about the weather, does so for very good reasons. What they are doing is acknowledging the fact the weather can exert a direct influence on their activities. They are willing to pay to reduce the level of uncertainty and to buy themselves time to plan to overcome the deleterious aspects of the weather. This is no different from a company that exports much of its output needing to monitor information about exchange rates because it is crucial to know what is happening to the value of the currencies in the countries with which it deals. They cannot control the rates, as they cannot control the weather, but if they are able to correctly anticipate trends, they can plan accordingly. The only difference between a company that monitors exchange rates and one monitoring the weather is that one is looking at an economic external pressure and one at a physical external pressure. Which one is easier to forecast is a matter of debate!

Political pressures

One of the most important influences which result in pressures being exerted from outside on a business are those resulting from political processes. The political process results in legislation and the interpretation and implementation of that legislation. Part of the importance of this area comes from the numerous players and the complexity of the process by which particular interest groups, of which there are many, get their message across (see Figure 4.4).

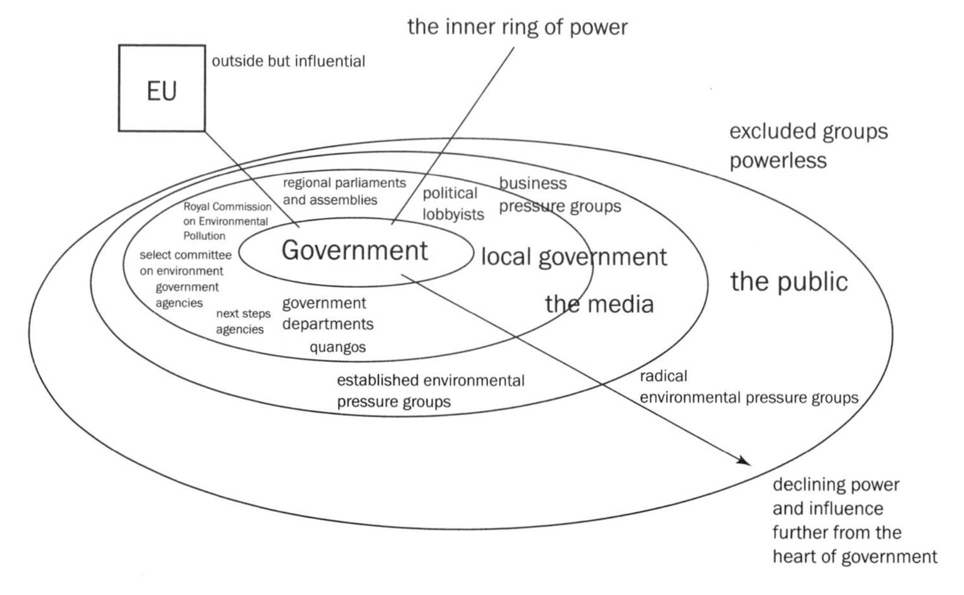

Figure 4.4 ***Power and effectiveness in influencing decision-making***

Parliament passes legislation and formulates the legislation, which determines the framework of environmental law. However, the reality of the process is considerably more complex than at first appears. The European institutions of the European Commission, the Council of Ministers and the European Parliament are having an increasing influence on the UK legislative processes. The Commission, headed by its twenty commissioners, implements the policies of the EU. It enforces laws and is the driving force behind new environmental policies. Internally it is divided into a number of directorates of which DGXI (Directorate General 11) is responsible for the Environment including nuclear safety. The Council of Ministers is the high profile decision-making body which confirms national compliance to EC legislation. The European Parliament has become increasingly important in recent years, moving from a consultative to a policy-making body. The perception of the Parliament varies in different Member States, the differences largely being a function of how differing Member States look on the institutions of the EU. Since 1993 there has been a European Environment Agency (Bell, 1997) which collects information and informs other European institutions on relevant matters. In future this agency may take on a more central role. Despite the increasing influence of the European institutions, most European legislation still requires enabling legislation to be passed through the Houses of Parliament for it to become part of UK law.

The UK Government is a complex organization and many of the government's powers in the area of environmental legislation are delegated to other bodies within government, or around the fringes of government. The government is made up of a number of departments, of which a surprising number have an influence on environmental policy and legislation. Often this causes conflict. For example, the Ministry of Agriculture, Fisheries and Food (MAFF) gives support for intensification of farming practices, while the Department of Environment, Transport and the Regions (DETR) gives support to mitigate against the effects of farming intensification. The Department of Environment, Transport and the Regions is the most important, but far from the only department, with an environmental brief. This department fulfils a number of different roles. It is the government department with overall responsibility for environmental protection and so has both a custodial as well as policing role in this area. It has authority over a number of agencies including the Environment Agency, the Countryside Agency and English Nature, who between them implement a number of policies which are determined by the DETR. The agencies report to and receive their budgets through the DETR. The Department also has responsibility for regional and planning matters and for local government. Therefore many of the environmental responsibilities handed down to local government can be traced back to the DETR. In addition, the department, as its name suggests, has responsibility for transport policy where environmental considerations form a significant part of the brief.

However, the DETR is not the only government department with which we should be concerned. The Ministry of Agriculture, Fisheries and Food and the Department of Trade and Industry both have considerable impact over the subject matter of this book. There is ample opportunity for conflict of interests here. DETR are charged with defending the environment while the ethos of both MAFF and the Department of Trade and Industry (DTI) (which now has taken on the responsibilities of the old Department of Energy) tend towards exploiting the environment. We should be aware of the role of other departments. For example, the Treasury has considerable influence over the spending departments we have described. The Department of Culture, Media and Sport has responsibility for both a number of sporting bodies and for English Heritage, who are the custodians of much of the architectural and cultural heritage of the nation. Many responsibilities for environmental matters in Scotland, Wales and the Northern Ireland have traditionally been handled by the three respective Secretaries of State. Now many environmental responsibilities are handled by the Welsh Assembly, the Scottish Parliament or the Northern Ireland Assembly. Throughout the difficult recent political history of Northern Ireland many environmental matters have been handled locally and since the environment is perceived as a non-political area, there are cross-border institutions which have operated successfully for a number of years.

In addition to the national government there is also local government, consisting of district and borough councils, county councils and, since 1992, unitary authorities. Most of these bodies' responsibilities, at least in the areas with which we are concerned, require the various councils to report to the DETR. Local government has responsibility for planning and land zoning (although some of these responsibilities vary in National Parks), for aspects of public health, for waste collection and disposal (of which there are several categories including, domestic, trade and hazardous waste), for some road and transport matters, noise and the enforcement of the Clean Air Acts.

The role of the government is completed by a number of agencies that implement the policies determined by the various government departments. The most important of these, as far as we are concerned, is the Environment Agency (EA). This was set up in 1995 as an independent corporate body created to ensure integrated pollution control. What it does is perhaps best understood by knowing that it was formed by a merger of the National Rivers Authority (NRA), HM Inspectorate of Pollution and the Waste Regulation Authorities. It therefore brought under one umbrella responsibility for the three major areas of waste disposal where pollution is inevitably likely to occur. For a full understanding of the responsibilities of the agency the account in Bell (1997) is perhaps the most comprehensive.

The second group of agencies of which the reader should be aware is those with responsibility for conservation and recreation in the UK. English Nature is the agency responsible for nature conservation in England. English Nature was the body that emerged from the old Nature Conservancy Council. It is charged with responsibility for the establishment and custody of National Nature Reserves (NNRs) and sites of special scientific interest (SSSIs). In England responsibility for recreation and amenity interests rests with the Countryside Agency. In Scotland and Wales the two areas of responsibility are handled by single bodies, Scottish Natural Heritage and the Countryside Council for Wales, respectively. The reason for the existence of two separate agencies in England has always been a matter of controversy since the present organization was instituted in 1991. Some people maintain that the emasculation of the old Nature Conservancy Council, which had been in existence since the National Parks and Access to the Countryside Act of 1949, came as a punishment for their over-vigorous defence of the countryside and the embarrassment they had caused the government. This vigour had shown itself most openly in the way the Nature Conservancy Council had sought to enforce the 1981 Wildlife and Countryside Act, which continued even after the Act was modified in 1985. Some of the detail of these machinations has been detailed in MacEwen and MacEwen (1987) and by Marren (1993). Within the National Park boundaries the park authorities take on many of the conservation and amenity responsibilities as well as being the statutory planning authority, at least in some of the National Parks.

It is sometimes difficult to be sure which agencies are part of the government and which are non-governmental organizations (NGOs). This is in part due to their complex and evolving history. For example, the waste regulatory part of the Environment Agency started out as being part of government, in this case the Department of the Environment (as it once was called). However, that part of the agency which had been the National Rivers Authority, prior to the privatization of the water industry, had been part of the Regional Water Authorities. The idea behind the creation of the so-called 'next step' agencies, of which the Environment Agency is one, is that they are to become semi-autonomous from the departments from which they sprang as soon as possible. Other bodies, such as the Forestry Commission, have also distanced their links and responsibility to the government as part of a general policy of deregulation. At the same time other regulators such as OFWAT (the Office of Water Services) have been created to regulate the new independent agencies and authorities.

The setting and implementation of legislation are only one half of the political process. There are a whole series of organizations that wish to influence the political process and to get over a particular point of view. The literature (Lowe and Goyder, 1983) distinguishes between interest and principle groups. So the Trades Union Congress (TUC), the Confederation of British Industry (CBI) and the Institute of Directors (IoD) would all fall within the first group. The latter do not represent a particular sectional interest but, rather, represent a set of values. The National Council for Civil Liberties or Friends of the Earth are good examples of such a group. Within this framework of principle and interest groups Lowe and Goyder (1983) describe two other groups which fit within the principle group. First, we have emphasis groups, who since their main aim is to defend existing standards, can be seen as preservation groups. This grouping tend to be made up of established organizations, which sometimes can acquire the status of semi-official watchdogs. Hence the Council for the Protection of Rural England (CPRE) or the Ramblers Association are consulted by the government on particular matters. The Soil Association is the body that awards the 'organic' status to individual farms. 'The National Trust has come to act as a *de facto* government agency' (Lowe and Goyder, 1983). For many such emphasis groups having the word 'Royal' as a prefix to their name is a clear sign they have been accepted as part of the establishment. Lowe and Goyder's second group is the promotional grouping: so-called since their objective is to promote social or political change. They tend to be more radical and closely aligned with campaigns to bring about change. Hence the campaigning groups Greenpeace and Friends of the Earth (FoE) fit into this category.

The various categories need not be regarded as mutually exclusive, however. The case of the Ramblers Association shows this clearly. The Association was originally set up as a walking club – clearly putting them in the interest group pigeon hole. However, they have come to represent a range of countryside interests almost in the spirit of the ecocentric nineteenth-century model. This places them as a principle group. They are consulted on footpath matters, most notably by local authority footpath officers, and are therefore seen as an emphasis group, defending the existing footpath network. At the same time they have acted as a radical promotional group in relation to access to the countryside. This was most obviously demonstrated in their connection with the mass trespassers of the 1930s, when walks were arranged across private land to which access was denied to the general public, most notably on Kinder Scout in the Peak District (Hill, 1980).

The various groups fulfil two important roles. First, they act to influence the political process either by direct lobbying of parliamentary or government interests and second, by raising public awareness of particular issues. This is done in the expectation that the public and particularly the press will be able to influence the political process. There is evidence that interest groups of various hues have had an influence on government policy. This has resulted in the government being forced to act quickly once issues have come to public attention. The case for legislation to acquiesce to public concern (or to ensure public health standards) in both the BSE crisis of the early 1990s and the concern over genetically modified (GM) foods of the late 1990s both demonstrate this point.

A feature of the political process in most Western countries is the extent to which interest groups wish to influence the political process by political lobbying. All interested parties in any area of public concern wish to gain access to the decision-makers. This can be achieved through their own representatives having the ear of those in high places, through the services of public relations consultants or through political lobbyists. On the industrial front, the CBI, the IoD, the TUC, along with various trade organizations, including such bodies as the National Farmers Union (NFU), the Road Haulage Association and the trade unions, all have MPs representing their views and gaining access to the political process. Many of the more established environmental groups are similarly placed. The National Trust, the RSPB, and the Worldwide Fund for Nature (WWF) have individuals willing to advance their cause. Even once radical groups such as FoE or Transport 2000 are able to have their voice heard in the corridors of power, while local groups can often call on the assistance of their local MP.

No consideration of the political process would be complete without acknowledging the role of the media. The media which includes public broadcasting, the serious press, specialist environmental, business and trade publications and magazines fulfil a number of roles. They provide a voice for the work of the various groups. The media provide two-way communication with government. Not only are they important as a conduit for interest groups to raise their particular agenda with government, they also provide an important way for the government to articulate policy and intentions to the general public. For established groups the media provide an alternate way of getting their message across to politicians. For more radical groups the newsworthiness of direct action is the main way for their cause to gain public awareness and so have some influence on the political process. Often the media themselves publicize particular events through a number of consumer and current affairs programmes and through the tried and tested techniques of investigative journalism. The interchange between media and government is a two-way process. Ever aware of the need to have their side of any issue aired by the media, the government is always keen to publicize their particular viewpoint on any environmental matter.

Environmental legislation

One of the main sources of pressure on business comes from the implementation and interpretation of current legislation. Not only does legislation limit the freedom of companies to act in ways they might choose, given a free choice, it also results in major conflicts with the environmental movement. Such conflicts are central to the themes developed in this book. Although it is outside our brief to examine the current legislation in detail – a task much better carried out by others, notably Bell (1997) – it is useful to be able to appreciate the broad fields covered by the legislation.

There are three major areas of legislation relevant to the environment of which business must be aware. The first relates to planning and the fundamentals of land use and land zoning. Central to this are the various Town and Country Planning Acts dating back to 1947 (with many subsequent amendments). This is the legislation which initiated structure plans, which outline the strategic land use issues and local plans which are more detailed and identify particular land uses and zoning for individual parcels of land. The Acts determine where major industrial developments are permitted to be located. They also established the Green Belts around our major conurbations and determine the balance of developments between greenfield and brownfield sites. They have a major influence on where particular activities take place, from the broad county level to the very detailed and very specific local level. They therefore have a clear influence on where business can operate and the resulting landscape of the countryside and urban areas.

The second major category of legislation is that of pollution control. This is the area that seeks to limit industrial pollution and the field with which many books about business and the environment have been almost exclusively concerned. Legislation in recent years has undergone considerable change due to a number of important developments. First, there has been a move to a more integrated approach to pollution control. This means that controls of emissions into the air, fresh and estuarine waters and the ground are now incorporated into a single structure. It might be argued that this has come to the UK rather late in the day, having been enshrined in the legislation of the USA and most Western European countries for many years. The second development is for UK legislation to adopt an approach more in keeping with that of our European partners, something which has brought about considerable changes to the UK system. The third significant shift is the adoption of economic instruments for the enforcement of legislation. This has resulted in the Polluter Pays Principle (PPP), being at the heart of much of the legislation, as we shall see in the next section of this chapter. The 1990 Environmental Protection Act is now the cornerstone of UK legislation, dealing with, among other things, the main forms of industrial chemical pollution. It consolidates existing legislation dealing with the various sources of pollution of the air, land and water and aligns UK legislation with its European treaty obligations. It was this legislation which led to the creation of the Environment Agency. There are two principal parts to the legislation. First, it identifies a substantial number of proscribed substances with the aim of preventing, or at best limiting, their discharge into the environment. The Act identifies what can broadly be described as the main sources of industrial pollution. Second, it seeks to assess the impact on the environment of all industrial emissions and then to control them.

The third major area of legislation relates to the conservation and protection of wildlife and the natural environment. Two pieces of legislation are central to our study. The first dates back over half a century – the 1949 National Parks and Access to the Countryside Act. It was this legislation which not only established the National Parks of England and Wales, but also the Nature Conservancy (which in later incarnations became the Nature Conservancy Council and then English Nature). This body has been given the task of establishing and managing SSSIs and National Nature Reserves, along with a variety of other responsibilities. The second central piece of legislation is the 1981 Wildlife and Countryside Act. This Act established a mechanism for the protection of individual species, for specific habitats and for the place of footpaths in the UK countryside. Both these pieces of legislation have been under almost continual bombardment from their critics. This is because they are in the front line between the pressures of business to exploit the land, on the one hand, and the perceived need to conserve whatever is left, on the other.

While we have identified some major areas of statutory law, it should be remembered that there are numerous other pieces of legislation that are relevant to our case. The common law should not be forgotten, especially where it relates to nuisance and trespass. It is worth noting, however, that this is more usually applied to individuals, since the legal complexities are much greater when applied to companies. There are numerous elements of company law which are relevant to how business interacts with the environment. Particular environmental businesses, for example, forestry, agriculture or those associated with tourism are all covered by separate legislation as well as by the planning, pollution and conservation statutes.

Legislative enforcement and interpretation

While environmental legislation may be passed with the best of intentions, the way in which it is enforced and interpreted presents a particular set of problems. Some of the most fundamental areas of environmental degradation, particularly those relating to the whole world, present the most serious problems for enforcement. International law is most difficult to enforce effectively. While the establishment of the International Court of Justice in The Hague has gone some way to establishing a court with international standing, it is difficult to bring cases without government backing and governments are unwilling to forfeit their national sovereignty. This is made more difficult by the fact that it is nation-states or representatives of those states who are likely to be in the dock. While there have been some successes, such as the IMARCO legislation (under the International Law of the Sea, which relates to pollution at sea from tankers), such examples are rare. The commitments of nation-states given at the Rio Agreement in 1992 have not been universally met and there are numerous examples where it has proved difficult to enforce international law. For example, Norway and Japan have both refused to comply with their obligations to the International Whaling Convention to cease commercial whaling operations. What sanctions are open to, for example, the United Kingdom, to rectify this position? Neither expelling all Japanese industrial investment from the UK nor declaring war on Norway seem viable options!

The way in which EU law has been adopted in the UK is a further example of tension emanating from the different approaches taken by the UK and her European partners to a similar set of problems. Environmental protection in the UK has traditionally been based on the discretionary flexible approach using the 'best practicable means' (see below). In contrast, European legislation has been based on the enforcement of statutory limits. The picture in Europe is further complicated by a culture of different states wishing to move at different rates along the road of environmental protection. Klatte (1991) has described a hierarchy of varying environmental zeal between the Member States of the EU. At the top of the table was The Netherlands, Denmark and Germany (to which Sweden should now be added), and at the bottom Portugal, Spain and Greece. He regarded the UK as reluctantly implementing European approaches. The scope for reluctance is most certainly present. Only European regulations are binding on Member States whereas Directives are translated into national law and enforced by the prevailing national standards. Since the UK has traditionally taken a different line in this area to most of Europe, there is no framework into which the new standards can be slotted and the transformation is proving difficult and sometimes expensive for the UK to adopt. A further complexity comes from the fact that the EU has developed a variety of policies from its different Directorates which, having allowed for subsidiarity and the different national interpretations, can sometimes lead to contradictions. So policies relating to the implementation of the Common Agricultural Policy (CAP), of the environmental policy, of the social chapter and of the policy towards the regions may produce unclear final objectives.

Enshrined in much of UK legislation (at least up until the Environmental Protection Act, 1990) is that of the 'best practicable means' (BPM). The concept is seen as encapsulating the discretionary approach to environmental standards, which has characterized UK legislation, since it was first used in the 1874 Alkali Act, although it has never been defined as a statute. It has been a matter of negotiation between the factory owner and the Alkali Inspector as to what was the best practicable means for minimizing the effect of the pollution to acceptable levels which has led to suggestions that the regulators were being too soft on the polluters. It was pointed out that most of the Inspectorate had been recruited from individuals who had once worked in the polluting industries. The comparatively small number of prosecutions suggested to critics that the poachers turned gamekeepers were not as rigorous as they might be. The environmental lobby asked whether the best practicable means was the best for the polluters or for society. The Alkali Inspectorate was not the only agency where there was disquiet concerning the cosy relationship between polluter and regulator. In the late 1980s the National Rivers Authority was the water regulatory authority watching over the polluters

– the most obvious of which were the newly privatized water undertakings. The two groups had been colleagues in the Water Authorities prior to privatization. While there was considerable zeal on the part of the field staff of the new NRA to enforce pollution limits, there was reluctance to do so at higher levels. Political pressure was brought to bear to protect the fledgling water companies from a spate of high profile prosecutions.

By 1990, with the impending introduction of the Environmental Protection Act and the subsequent creation of the Environment Agency, there came about a considerable shift in thinking. The concept of BPM was replaced by the inelegant acronym BATNEEC (best available techniques not entailing excessive cost). Although still representing a trade-off between commercial and environmental objectives, the 1990 legislation should be seen as a move towards integrative pollution control. It was to move at least halfway towards the European model for environmental control. Through the various regulations, much of the discretion of individual inspectors was removed. A further development of the 1990 legislation was that polluters had to take into account the best practicable environmental option. They are not obliged to adopt this option, but, by virtue of the fact they had to consider such an option, it meant they were now, in effect, obliged to carry out an environmental audit of their operations.

Environmental legislators have been dogged by a simple problem: the solution to which is rather more difficult to achieve. Most legislation's principal aim is to reduce the impact on the environment of a particular entity. For example, it may be to reduce the amount of sulphur dioxide present in the atmosphere. The only effective way in which this can be done, however, is to limit the amount of sulphur dioxide emitted into the air by individual polluters. The majority of polluters only make a minuscule contribution to the overall problem. While the emissions of each individual can be limited, it is difficult to restrict the total number of polluters. Attempts to control emissions of a range of pollutants, into the air, into water courses, into the ground, from ionizing radiation, noise, etc. have resulted in a variety of differing ways in which regulations can be framed.

There are basically two groups of standards that can be identified:

- standards determined at the source of the pollution;
- environmental quality standards.

Many source standards are expressed in terms of the emission of various substances, where either the maximum concentration of the pollutant is stipulated, or where the total amount of the pollutant discharged over some specified period of time is laid down. Other standards of this group may be in terms of the process that must be adopted. Sometimes the process is specified, or sometimes the composition of the raw materials is agreed, for example, it may be stipulated that a particular quality of oil has to be burned in a furnace, or the conditions of operation may be prescribed. This means that the particular way the plant is operated is specified. Product standards are sometimes employed in that the equipment must reach certain standards and the equipment may have to be frequently checked, for example, road vehicle emissions are checked at MoT tests to ensure they reach particular standards under test conditions (which does not guarantee they reach the same standards at all times they are in use). Sometimes design standards are employed so the equipment in use must reach certain design criteria when new, for example, solid fuel boilers in smokeless zones have to meet certain design criteria and cars have to meet basic safety criteria before they can be sold to the public. In most cases the standards are established in such a way that they offer a practical approach to ensuring the desired standards are achieved. Sometimes target standards appear to be the way forward. These specify not how much pollution is created but how much reaches the 'victim'. These may be based either on a level which has an adverse effect on humans or which results in a long-term detectable biological degradation on the environment.

When standards have been fixed and the legislative process completed, the various laws and regulations still have to be enforced by regulators and complied with by business. Attitudes to breaching regulations have changed and there has been a gradual shift from polluters being able to talk themselves out of a tight corner, to one where there are penalties for non-compliance. Even today there are criticisms of the low level of prosecutions and what many see as the derisory levels of fines imposed on polluters. Many argue these mean pollution incidents are likely to continue. However, there has been a shift in both regulators' attitude and that of society at large. Clear guidelines and regulations after 1991 replaced many of the discretionary powers of the Environment Agency's Inspectorate. There has been a shift in thinking away from an attitude that this was a breach of some administrative regulation, which could have happened to anybody. Once pollution incidents were regarded at worst as a victimless crime. Now the penalties are worth avoiding. The range of penalties has also increased in more recent legislation. These may take the form of administrative sanctions from the regulators to suspension or modification of the discharge licence.

Polluters now have to meet clean-up costs, so where the local authority, or another agency, is involved in the clean-up and disposal of an incident, the costs can be recovered. Companies, or individuals in companies, may be subject to criminal charges. This is important since it demonstrates a real crime has been committed. While the criminal law has been strengthened, individuals or organizations still have recourse to civil action in the courts to redress environmental wrongs, although this course of action can prove expensive. For many companies the adverse publicity resulting from a court appearance has a salutary effect on their activities. Arguably, this is more effective than the level of fines. The naming and shaming of companies in lists of polluters by either environmental pressure groups or the regulators are now having an impact. Many firms will go some way to avoid being included in such lists.

Economic pressures

Since this book is not a text on environmental economics and since we covered some of the more important economic ideas in Chapter 3, we give here only a brief outline of some of the main concepts that touch upon business and environmental issues. We recommend those in search of more detailed accounts of environmental economics consult the works listed in Further reading at the end of the chapter. However, the economic pressures that come to bear on a business require an overall understanding of basic economic principles. Turner, Pearce and Batemen (1994) set out to provide 'a general audience [with] the basic concepts and principles of what has become known as environmental economics' – a task which they managed admirably. Pearce and Turner had previously (1990) written *Economics of Natural Resources and the Environment*, which is widely acclaimed to be a comprehensive text on the subject but one which may only be fully appreciated by what they themselves term 'mainstream economics undergraduates'. Tietenberg (2000) provides a very good introduction to environmental economics, which is comprehensible to those unfamiliar with economic terminology.

Businesses are always aware of economic factors that may come to bear on their activities. Where there is economic growth, as expressed by a combination of the business cycle, GNP trends and disposable income, then there is a greater likelihood of environmental concern coming to the fore. During periods of recession, keeping a job and maintaining a standard of living become more important than environmental concerns for most people and for most businesses. The different economic climate in individual countries means that there is differential environmental concern between states. Levels of taxation are a further economic factor that can lead to environmental pressure. There is increasing awareness of the possibility of the introduction of a carbon tax that would make businesses reappraise their present practices and look for alternatives to carbon-based fuels.

Governments have now found that market mechanisms are a popular way to achieve environmental objectives. This means economic rather than legislative or administrative procedures are increasingly used as a way of implementing environmental policy. With deregulation and with changing attitudes to the environment, economic market instruments represent a major tool to put pressure on business to comply with environmental standards. These are based on the concept of the polluter pays principle (PPP) which has been popular, since at least the 1970s as an incentive rather than a cure to environmental problems. It emphasizes that the total costs of a cure, not just the direct costs of clean-up, will be borne by the polluter. However, determination of the exact true costs are often difficult to ascertain and to attribute as we have seen in Chapter 3. Bell (1997) lists a number of ways in which market mechanisms have been implemented. First, charges may be made for the administrative costs involved in environmental inspection, licence granting and monitoring rather than these costs being borne by the taxpayer. So consents for the discharge of waste water are charged to companies. While a company may see this as paying twice – once for permission to discharge and then, second, for the discharge itself, this reflects the total costs involved. The second popular market mechanism is for taxes to be levied to reflect the total environmental costs. Often attributing such costs presents a number of difficulties. For example, to take into account all the environmental costs involved in the use of heavy lorries on the roads and then to arrive at an excise duty for such vehicles is a very complex operation. Costs involved in pollution prevention or remedying particular incidents are a third market mechanism. These are well developed for water pollution where licensing arrangements are well advanced. Sources of water pollution are often discrete and therefore can be accurately pinpointed. Air pollution presents a more difficult challenge for such an approach since it is much harder to identify culprits accurately. Air pollution can arise from many different sources and the rapid mixing of pollutants in the atmosphere and their transport by the wind renders the accurate identification of the originators problematic.

All the above examples can broadly be categorized as attempting to internalize the total cost of any real or potential pollution to those individual companies who are likely to have caused it. There is also another range of market mechanisms available. Subsidies and grants may be used to encourage particular policies as has happened with farm management grants. There has been an increasing tendency for quotas for pollution to be traded. Hence where one polluter, be it an individual or a country, pollutes less than its quota, it may trade (for any tradable commodity including cash) with another party who is willing to pay the asking price to use the other party's quota. While the ethics of such action is questionable and while it does not result in a net decrease in pollution, it does highlight the real cost of excessive pollution. The final market mechanism, which has been extensively used, is that of various deposit and refund schemes. Legislation now exists in the UK to get obligated businesses to recover materials as part of their operations. Whether they have done this is demonstrated by obtaining packaging recovery notes (in which a trade has developed). These schemes have not been fully adopted and may yet result in numerous administrative problems. Similar plans for Denmark to recycle bottles have been shown in the courts to breach free trade agreements and schemes to recycle drinks containers in Vermont, USA, in the 1970s, resulted in administrative chaos. In the UK in the 1970s the idea of recycling drinks containers proved impractical in the light of opposition from major supermarkets at that time.

Sociocultural pressures

The sociocultural pressures exerted on business arise from two sets of factors: what people do and what people think. What people do is seek greater well-being and a better lifestyle. Leisure time has increased and working hours decreased for many individuals, although some of this has been eaten away by increased commuting time, particularly in metropolitan areas. In addition, people are living longer and so overall the provision for recreational activities has increased significantly. Increased wealth has allowed people to buy their way into a better lifestyle by securing either goods or services, i.e. by consumerism. These activities invariably use up resources and in turn produce pollution. Often they result in finished products that ultimately have to be discarded. The twentieth century saw fundamental changes in lifestyles. There are now social pressures to visit the countryside, to indulge in leisure pursuits and this in turn leads to businesses setting up in such areas to meet the demands. Provision for tourism, activity holidays, better roads, communications and infrastructure along with the manufacture of a range of leisure products from deck chairs to powerboats are seen as a manifestation of the response of business from both the manufacturing and service sectors.

At the same time – and largely conflicting with these trends – is superimposed a second set of sociocultural pressures emanating from individuals' attitudes and opinions. Most individuals believe that environmental improvement, conservation of resources and their heritage, along with moves to sustainable futures, are worthy things with which to be associated. These beliefs are reinforced by exposure to education – where the young are particularly receptive, from media exposure and from the subtle and not so subtle protestations of the environmental movement. These social pressures have forced businesses to be more environmentally aware but at the same time to see that such thinking may offer considerable business opportunities. The growth in the market for organic food offers such a possibility, with a price premium for the producer but more particularly for the vendor.

Technological pressures

At the heart of the debate about the way in which technological factors put pressure on business is a conflict between those who take a technocentric approach (believing in the possibility of a technical fix to our environmental problems) and those who have a basic lack of confidence in science or at least our ability to apply that knowledge. Technological pressures basically come in the form of pressures to adopt new technologies. In the last fifty years there have been a number of significant scientific discoveries, coupled with an increase in the speed of technological transfer, which together have generated pressures to move on from obsolete technologies. Discoveries can be classified as:

- mechanical/physical – new manufacturing processes;
- chemical – increased use of hydrocarbons (plastics) and organics (pesticides and herbicides);

- biological – genetic engineering and complex biotechnical compounds.

The information technology revolution is a fourth category of development that depends on an interaction between hardware and software advances. There is often a considerable lag between scientific discovery and technological adoption with the technological pressures coming from the perceived need to adopt the technologies. For example, the motor car was 'discovered' a century ago although the major pressures it has been responsible for generating have come within the last thirty years.

There is a range of technological pressures associated with obsolescence. Social pressures to do things in 'the new high tech way', produce short product life cycles. Legislation to comply with new technical standards and a move to adopt bigger and more powerful machines, brought on by the increased speed of technological transfer, reinforce pressures to believe in the technological fix. This leaves a large pile of obsolete, rusting equipment, which is difficult to reuse or recycle.

While there are undoubted pressures to adopt new technology, there is also a deeply held mistrust in the new technology. The revelations about pesticide pollution – unearthed by Rachel Carson in *Silent Spring* (1962), Chernobyl, pollution incidents such as the *Exxon Valdez* and *Amoco Cadiz* disasters, revelations about the impact of greenhouse gases and the destruction of the ozone layer have all pointed to sustainable technologies being the way forward. There are therefore technological pressures now to develop sustainable technologies. Many of these are now on the point where they are economically competitive with technocentric solutions.

Environmental impact assessment

One of the main pressures business has been subjected to in recent years comes, not from what is happening today, but rather, from the impact today's business is going to have on the environment in the future. Businesses now have to be prepared to make a clear statement on their plans for the future and of their likely impacts. Some businesses see this as another case of administrative interference on their activities. To others it is a helpful adjunct to their internal planning processes. This requires a business to evaluate what it is doing in terms of the total impact on the environment. We finish this chapter by examining an important predictive tool: the environmental impact assessment (EIA) used by companies, legislators and increasingly environmentalists, to evaluate the likely effects of proposed developments on the environment. A series of methods has been developed to carry out these assessments. They have become more sophisticated in the last decade and are now gaining more widespread acceptance.

The basic premise behind all environmental impact assessment is that a business is required, as part of the planning process, to evaluate the total impact that a project is likely to have on the environment. This may result in a formal environmental impact statement (EIS). What is laid out in such a statement is usually laid down in the relevant legislation. Some companies may decide to prepare such a statement even if not required to do so by law. Although there are variants most schemes require proposers to do the following:

- catalogue all the environmental effects of the project;
- analyse the relationships with the relevant ecosystems and social processes;
- evaluate all the residual consequences of the project.

By doing these things a debate is prompted on the likely environmental consequences of the project and forces all the interested parties to evaluate the various pieces of evidence available. This results in a more objective evaluation of the evidence than would otherwise be the case, and means the total range of impacts are open for discussion. It avoids the more subjective nature of submissions at the formal planning stage that have dogged many planning inquiries in the past.

Most of the current thinking on EIA came out of the USA as a result of the National Environmental Policy Act (NEPA), which became law in 1970 (Andrews, 1976). This Act required proposers to submit environmental impact statements for any significant projects. The law enabled environmental groups to force federal agencies, among others, to produce EISs. The law was policed by the Council for Environmental Quality and although it did much in its early years to force proposers to address environmental issues, it was also regarded as a tool which slowed down the development process. The legislation was subsequently modified in the USA and arguably, due to the administrative and legislative burdens it imposed on business, allowed objectors to block the adoption of similar legislation in Europe. In the 1970s when NEPA was being pursued with vigour in the USA, public enquiries in the UK were becoming large cumbersome affairs – particularly enquiries for large projects such as road schemes and nuclear power stations. These are just the sort of schemes that EIA should prove useful in evaluating. However, due to the fear that EIA would result in still further hurdles which proposers would have to negotiate, there was severe opposition to their incorporation into the planning process. Some proposers, however, saw it as a way of combating opposition under certain circumstances. It was argued that by demonstrating the environmental implications of a proposal had been considered, it might ultimately speed up the process. Hence British Gas used EIA in planning its gas terminals along the North Sea shores and the National Coal Board used the technique extensively in the planning stages of the development of the Vale of Belvior coalfield. There was large-scale opposition, however, to its widespread introduction. For example, MAFF refused to have any part in adopting EIA on the grounds that agricultural developments were outside the planning process and the entire Channel Tunnel project was conducted without any recourse to an environmental impact statement. This was despite the fact the European Directive 85/337 should have been adopted by all the Member States after July 1988 and that the Labour Government as early as 1978 had accepted that EIA might have merit for large schemes and nothing came any bigger than the Channel Tunnel.

The EC Directive 85/337 on the Assessment of the Effects of Certain Private and Public Projects on the Environment is the foundation for the need for environment impacts statements to be made under European environmental legislation. The UK is not alone in failing to implement the directive, even years after its original adoption date of 1988. UK legislation even now is not clear as to which projects require an EIS to be submitted as part of the planning process, nor exactly when such a statement should be submitted (Bell, 1997). The Town and Country Planning (Assessment of Environmental Effects) Regulations 1988 identify two schedules, based on the original Directive. Schedule One lists those projects where an EIS is mandatory. These include power stations, oil refineries, nuclear waste facilities, heavy chemical works and motorways. Schedule Two originally intended that an environmental impact statement would only be mandatory if it was considered that significant environmental effects were likely. The list in Schedule Two includes paper mills, food processing plants, glass works, mineral extraction sites, pig/poultry units and, interestingly, holiday villages. Since the government has accepted the argument that agricultural projects do not need planning permission and has been reluctant to make a decision as to which projects would make significant environmental impacts, the mandatory element of this schedule has never been tested. The original Directive used the term 'project' when referring to when an EIS was required. This has been translated in the UK into 'development' as defined in the planning legislation. It has therefore been argued that where an 'undertaking' does not require planning permission, it cannot be regarded as a 'development' and therefore does need to provide an EIS.

The process of carrying out an EIS can be approached in a number of ways. Even today many assessments are carried out by use of a matrix listing the possible impacts. Leopold *et al.* (1971) first used this technique. Leopold arranged a matrix consisting of a horizontal list of development activities and a list of environmental characteristics along the vertical. For each cell in the matrix two scores are entered, one for the magnitude of the impact along with one for the importance of the impact. Both scores are ordinal values within the range of 1 to 10. Although there are a large number of potential cells for any project, the impact will be concentrated around a limited number of cells, which tend to be bunched around particular areas. With the passage of time more sophisticated variants of the idea have evolved. Sophisticated overlays using geographical information systems (GIS) have been developed together with computer simulation models to predict the various impacts.

We must see the present situation as one in which business needs to realize that with increasing environment awareness, accountability for future impacts of projects will increase. The present regulations are complex and only apply to major projects. Although they are carried out objectively, the same cannot be said for how they are evaluated as part of the political process. They are implemented with varying vigour and at different speeds in different parts of the world. Europe has taken on the basic concept but managed to avoid the administrative nightmare of the early days of NEPA in the USA, which threatened to delay any development almost indefinitely. It is highly likely that business will be subject to increasing pressures in the future to produce some form of EIA for any significant development.

Summary

Chapter 4 has shown that businesses are subject to many external influences over which they have little or no control. There are a whole range of such influences arising from the workings of the economy, society, technological developments and the political system. Of the total range of external factors that influence business activity only a small, but important, proportion could be described as environmental. Some of these are brought about by the activities of other businesses, some would have occurred even if no other business existed. The natural environment and environmentalism influence business in many ways. These can be categorized in terms of the commonly used PEST analysis, although this may be seen as a simplification of the way in which many of the influences are linked together. In addition, we have identified the physical environment as having a direct influence on business activity. An indicator of the growing significance of external influences has been the development of a host of regulations relating to business and the environment. Reaction to external influences is one way of dealing with pressures, but as we shall see in Chapter 5, businesses are increasingly becoming pro-active in their attitudes to the environment.

Further reading

One of the main characteristics of the reading for this chapter is the number of different academic disciplines that have been drawn upon and the large numbers of texts that could be consulted in the various fields. For an understanding of the way in which the physical environment interacts with the human activities then A.S. Goudie, *The Human Impact on the Natural Environment* (5th edition, Oxford: Blackwell: 2000) provides a good coverage of the material. L. Owen and T. Unwin (eds), *Environmental Management: Readings and Case Studies* (Oxford: Blackwell, 1997) provides a number of readings, several of which impinge on the subject matter we have mentioned. I.G. Simmons, *Changing the Face of the Earth: Culture, Environment, History* (2nd edition, Oxford: Blackwell, 1996) takes a more 'human' approach to environmental change, while G. Goldsmith and A. Warren (eds), *Conservation in Progress* (Chichester: Wiley, 1993) take up many of the conservation management issues which are identified within the chapter.

Moving on to the way in which the external influences come into play S. Bell (1997) *Ball and Bell on Environmental Law* (4th edition, London: Blackstone) provides an excellent and current view of both legislation and of policies emanating from the legislation. This text gives a good overview of the way in which European legislation and institutions are being enshrined in UK legislation. Other texts which give a good view of policy issues are N. Haigh, *Manual of Environmental Policy: The EC and Britain* (Harlow: Longman, 1993) where again the European dimension is well covered, and W. Baumol and W. Oates, *The Theory of Environmental Policy* (Cambridge: Cambridge University Press, 1988) where a more theoretical approach to policy issues are taken. For an understanding of environmental economics the reader is faced with a bewildering choice. The best introductory text is T. Tietenberg, *Environment and Natural Resource Economics* (5th edition, Reading, MA: Addison-Wesley Longman, 2000). However, R.K. Turner, D. Pearce and I. Bateman, *Environmental Economics* (London: Harvester Wheatsheaf, 1994) also provides an excellent overview. For those with more familiarity with economics then D. Pearce and R.K. Turner, *Economics of Natural Resources and the Environment* (London: Harvester Wheatsheaf, 1990) and A.M. Hussein, *Principles of Environmental Economics* (London: Routledge, 2000) are useful texts. There are numerous readings in environmental economics with perhaps H. Folmer, H.L. Gabel and H. Opschoor, *Principles of Environmental and Resource Economics* (Aldershot: Edward Elgar, 1995) providing an approach closer to that we have taken than many other texts. For a detailed text on cost benefit analysis then N. Hanley and C.L. Spash, *Cost Benefit Analysis and the Environment* (Aldershot: Edward Elgar, 1993) covers the subject fully. However, we anticipate many of the more general texts on environmental economics cover CBA in sufficient detail for most readers of this text. For a consideration of environmental impact assessment then P. Walthern (ed.) *Environmental Impact Assessment* (London: Unwin and Hyman, 1988) provides a good background although J. Glasson, R. Therivel and A. Chadwick, *Introduction to Environmental Impact Assessment* (London: UCL Press, 1994) provides a good alternative.

Discussion points

1 Critically evaluate the chief differences in the way EIA has been used in the USA and the UK.

Exercises

1 Conduct an environmental PEST analysis for any business with which you are familiar.

2 Identify all the government departments and agencies with responsibility for the environment. Search for their websites and make a list of their policies and responsibilities towards business and the environment. You could extend this to a search of European Union websites, or do a comparison of policies with a selection of different countries.

5 Environmental business opportunities: business becomes pro-active

- How do environmental business opportunities arise?
- What could an opportunity mean for a business?
- How business sees opportunities
- Constraints to seizing opportunities
- Physical opportunities
- Legislative and planning opportunities
- Planning laws
- Case study: mushroom picking in Washington State
- Social opportunities
- Green consumers
- The technical fix route to opportunities

> The man with a new idea is a crank until the idea succeeds.
>
> (Mark Twain)

> [B]y adopting a pro-active stance, the management will be better placed to consider the environment, not as an additional cost burden but rather as an opportunity for gaining competitive advantage.
>
> (Blaza, 1992)

> Honduras is hoping to lure tourists with a taste for destruction by inviting them to see the aftermath of Hurricane Mitch. The minister for tourism, Norman Garcia, said yesterday that the country would capitalise on a surge of visitors who have come to witness the effects of October's deadly storm which killed more than 5,000 people in Honduras alone. The ministry wants to offer a 15-day tour which includes the chance to see the destruction, to help with reconstruction and to visit the standard tourist attractions.
>
> (*Guardian*, 12 January 1999)

> 'Tis an ill wind that blows nobody any good.
>
> (Anon)

There can hardly be a more literal illustration of the old saying about an ill wind providing an opportunity for gain than the Honduras Tourism Ministry planning tours of hurricane damage sites. Here is a dramatic case of trying to turn an environmental disaster into a business opportunity. It also illustrates the difficulties of separating pressures from opportunities. They could be likened to two sides of the same coin. What is a disaster for one is a golden opportunity for another. Hurricane Mitch in 1998 was an undoubted disaster for the Honduran banana growers, but was later seen as a potential benefit for the tourist industry. The same environmental attribute is viewed very differently by different businesses. Two different businesses can react differently to the same aspect of the environment. The essential difference between business pressures and business opportunities is that:

- Pressures reach out to the business from external sources. They are not sought out by the business. They are generally unwelcome and they necessitate action to protect the firm. They are regarded as negative elements.
- Opportunities are sought by business. A firm reaches out to the opportunities. They are generally welcome to the business and are a means to enhance the firm's activities. They are regarded as positive elements.

Although the actual environmental element may be the same, the way it impacts on different businesses may vary according to the particular circumstances of the time and the type of business. We can illustrate this by considering a range of responses by business to an environmental attribute. The important point to note here is that the same aspects of the environment could be looked at in very different ways. As an example let us consider how a mining company might deal with a worked out open-cast pit.

The firm has open to it several options. First, it may decide to do nothing with it. This costs nothing, except perhaps a loss of goodwill among the local community. If it is in a scenically attractive tourist area, it may attract adverse publicity. The company may not be aware of any potential use or it may leave the pit 'fallow' in the hope of future reworking.

Second, the firm may consider using it for another purpose, but decide not to do so, because it is financially unviable or because it generates fierce public opposition. For example, filling in the pit and restoring it to its original farmland condition would be prohibitively expensive and could never generate enough income to cover the costs. Alternatively, proposals to use it as a landfill site might be financially viable but would create a storm of opposition from local residents and the media. It could be sold to another company or organization, which may use it for another purpose, but this ends the original company's involvement and foregoes the possibility of taking up any opportunity in the future.

Third, the company may use it for an obvious, but unimaginative, purpose based on existing conditions. For example, it may backfill the pit with topsoil sufficient for agriculture, but not attempt to restore it to the original ground level. This might be financially viable and would generate little opposition. Numerous examples of this can be found, for example, Pitstone Chalk pits, Buckinghamshire or the wet pits used for fishing and water sports at Rickmansworth Aquadrome, Hertfordshire. These could be described as existing or self-evident opportunities, in that they do not require the company to make very much change in the nature of the resource.

Fourth, the business may decide to create a new opportunity. The pit is used as a framework for something new which enhances the environment. A good example of this is the use of old china clay pits near Falmouth in Cornwall as the site for the Eden Project. Here the site has been designed to house diverse exotic habitats, including tropical forest and desert ecosystems, under glass. This sort of project requires innovation and imagination to succeed.

Finally, the company may turn a pressure into an opportunity. An abandoned mine poses significant safety problems since it attracts adventurous children, who will be at risk from the derelict buildings, old machinery, flooded workings, subsidence, pits and shafts. The company could face litigation if any intruders were injured or killed, which would be both a potentially expensive liability as well as a public relations disaster. Therefore the firm could turn the site into an opportunity by converting it for instance, into an industrial heritage museum. It would then become a paying tourist attraction. There are increasing numbers of old industrial sites which have been turned into profitable undertakings, for example, Iron Bridge Gorge, Morwellham Quay and several former coal mines in South Wales.

How do environmental business opportunities arise?

A changed environmental condition may provide the business with an opportunity that can be viewed essentially as a gap in the market – a need or want that the business can satisfy. The changed environmental condition may come about from a variety of causes, which may be interrelated. Four main types of cause can be identified:

1 *Natural causes*: for example, meteorological events such as the Great Storm of 1987 in Southern England, which resulted in a huge number of trees being blown down which led to a big demand for chain saw operators.
2 *Technological change*: the ever faster improvement in personal computers, means rapid obsolescence for perfectly sound machines. Elkington and Hailes (1998) said that by 1999 there would be 100 million obsolete PCs world-wide. They mention three organizations, Bytes Twice, Cybercycle and Recycle IT, in the UK alone, which collect, recondition and resell old computers. In addition, Digital, IBM and Xerox have also set up systems to recycle or dispose safely of obsolete machines.
3 *Legislation*: the 1956 Clean Air Act resulted in smokeless zones and the end to open coal fires in many built-up areas. This stimulated the conversion of most UK homes to smokeless central heating. This was an opportunity for the sellers of central heating, although a severe blow to the industries dependent on coal.
4 *Changed public tastes or behaviour*: for example, concern about water quality and health led to a huge increase in the consumption of bottled water in the 1990s.

All these changes create new costs and/or new demands on businesses. Here we are concerned with those aspects we regard as opportunities (costs have been dealt with in Chapter 4).

What could an opportunity mean for a business?

Business responses can take the form of either creating new opportunities or turning costs into opportunities. One way of creating an opportunity would be to invent a new product to solve an environmental problem – such as a garden shredder to turn woody waste into compost to increase home garden recycling and reduce the demand on wetlands for garden peat. Another way would be to develop a new service, for example an environmental business consultancy to provide advice for businesses on how to comply with new regulations on waste reduction. Businesses can also turn costs into opportunities in a variety of ways. They could devise new ways of doing existing things to comply with new legislation or to satisfy new markets. For example, the requirement for information about compliance could be satisfied through an Internet-based environmental data service to provide businesses with cheap cost efficient information such as that operated by ENDS (1999; HTTP://www.ends.co.uk). The adoption of new technologies may offer environmental opportunities, for example, the introduction of water-efficient fixtures at the Hilton Hotel in Novi, Michigan saved 70 per cent of the estimated normal water use for a building of that size and the annual water and sewerage bills were cut by up to US$45,000 (von Weizacker *et al.*, 1998, pp. 194). What could be a burden can be become an advantage by looking at the issue as a potential opportunity rather than as a cost.

How businesses see opportunities

What influences a business to seize an opportunity or fail to spot it? Not all businesses respond to the same apparent stimulus. There is no automatic or Pavlovian reflex that dictates the business must seize an opportunity. The answer lies in the decision-making behaviour of the business. This is influenced by a host of external and internal factors. Figure 5.1 shows how the various factors are inter-related and demonstrates the alternative outcomes for decisions of an environmental business opportunity.

A business needs to be aware of the changes in its operating environment that can give rise to an opportunity. Four principal external factors need to be addressed. First, changes in the environment (physical, social, political), for example, a worsening air or water quality. Second, requirements for compliance to legislation (national, EU or international); this is especially important for TNCs who may have to deal with a variety of different regimes. Third, general business trends relating to the environment, for example, changing best practice for environmental audits, expected standards from customers, suppliers or investors. Fourth, the behaviour of competitors who may seize opportunities and thereby develop an advantage or who may be new entrants into an already crowded market place.

Information about all the factors is brought together according to the firm's information collection and analysis processes. How this information is used, however, will depend on the culture of the company and in particular on the values, attitudes, experience and motivations of the key decision-makers. Eden (1996) and Welford (1997) have examined the

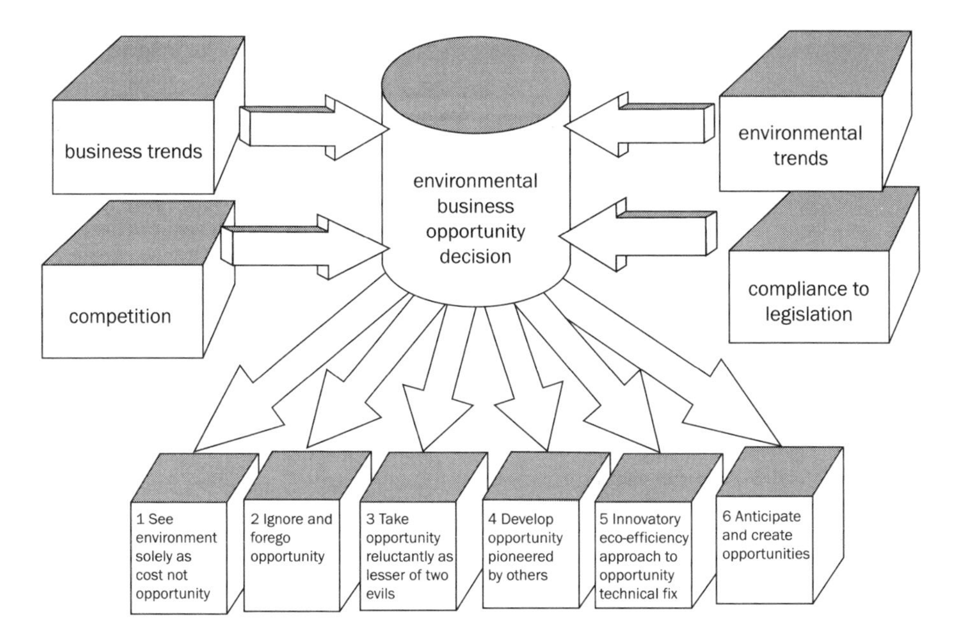

Figure 5.1 ***Alternative business decisions for environmental business opportunities***

influence of the company culture on decisions. Company culture affects all facets of the business not just environmental aspects. The leadership style of the company is vitally important in setting the tone of the company culture and a change in leadership may transform the decisions of the company. For example, Tesco was run in a very different way by its founder Jack Cohen and by his successor Iain McLauren (Wrigley 1988).

A business may approach environmental opportunities in a number of alternative ways. It might view the environment solely as a cost and not perceive any opportunity. It would therefore not take up an opportunity. Alternatively, it may recognize the opportunity as existing but decide not to take it up perhaps for cost reasons, or because it was averse to risk.

Sometimes a firm may take an opportunity reluctantly as the lesser of two evils. It may acquiesce to the opportunity without any real enthusiasm. This may come about as a result of compliance to legislation or of threats from competitors. A business could develop an opportunity belatedly but enthusiastically. Others will have pioneered the opportunity. This type of company does not seek risk and only adopts new ideas when proved safe by others. It is a follower of fashion not an innovator. Another firm may actively seek to develop opportunities as a cost savings measure or for revenue enhancement. It sees environmental opportunities as good business practice. It has an eco-efficiency motivation, using the environment as a means to improve the profitability of the business. Ultimately a company may develop into a sustainable business that seeks to create opportunities for both business and the environment. Although it tries to develop a profitable business its prime motivation is driven by the needs of the environment. Businesses such as this come in two sub-types. First, there are those which have been developed by non-profit organizations (such as charities, NGOs and quangos). They are mainly concerned with environmental goals and use their business arms to generate funds for their environmental objectives. The business enterprises of the National Trust and R S P B would fit into this category (see Box 5.1). Second, there are those businesses which are genuinely motivated by environmental goals but at the same time seek business goals. These are at present few and far between. Examples might include Interface carpets, Ecover, The Body Shop and SustainAbility.

Box 5.1

The National Trust's business enterprises

The National Trust is the UK's largest conservation charity with 272,659 hectares of countryside and 565 miles of coastline under its protection. It also owns over 1,000 ancient monuments, 165 historic homes, 19 castles, 234 landscaped gardens and parks, 700 working farms and 30 National Trust town shops. With 2.56 million members, it is the largest single environmental organization. Membership subscriptions accounted for just over 30 per cent of the Trust's £182.4 million income in 1998–99 and admission fees from non-members accounted for £8.6 million (4.7 per cent). National Trust (Enterprises) Ltd is a commercial arm of the Trust which raised £11.4 million (6.25%) income in 1998–99 through :

- mail order catalogues featuring NT-related products such as exclusive tableware and china with historic designs based on furnishings in NT properties;
- shops at properties selling both general NT products such as books, calendars, confectionery, toiletries, clothes and property-specific goods such as guide books, videos, postcards, scarves and tea towels with designs based on the site;
- town shops in historic locations not connected with NT-properties; they trade on the NT image of heritage and quality;
- restaurants at properties, some of which also cater for special events such as lecture lunches;
- holiday cottage lettings – many country house properties have extensive estate housing inherited from former times when they would have housed estate workers;
- National Trust Travel Collection – escorted holidays with heritage themes.

Source: National Trust, 1998, 1999.

The role of individuals in companies can be crucial. A committed company chief executive can direct a transformation, for example, Ray Anderson of Interface who became converted to the philosophy of sustainable business after reading Paul Hawken's (1993) *The Ecology of Commerce* (Armstrong, 1999). A suggested framework of attitudes showing characteristics and goals is shown in Table 5.1. The first three, evangelist, convert and technocrat are the most likely to be active in moving a company towards seizing environmental opportunities; the agnostic may be persuaded to become a convert, but the sceptics and the unbeliever are unlikely to be moved. They will only view the environment as a cost, never as an opportunity. Opportunities can be very diverse and it is useful to examine the relationship between the environment objectives and business objectives in the context of some examples to give an idea of the possible range. Table 5.2 shows the kind of opportunities associated with specific environmental objectives. These are of course not the only opportunities, merely a representative sample.

Table 5.1 ***Business attitudes to the environment***

Type	*Characteristics*	*Goals*
evangelist	knowledgeable about environment seeks to promote and convert others puts ideas into practice	sustainable business
convert	less knowledgeable and has not yet put ideas into practice but philosophically persuaded	moving towards sustainable business
technocrat	eco-efficiency as good business practice, but does not believe it necessary to convert whole business	eco-efficiency
agnostic	unsure if there is anything to worry about or not; listens to arguments; uncommitted	continues as usual
sceptic	not convinced by environmental arguments	business as usual
unbeliever	anti-environment; sees environment and environmentalists as a threat	*laissez-faire*

Note: attitudes may be held by individuals or companies
Predominant company attitudes may change suddenly due to changes in senior management.

Table 5.2 ***Environmental objectives and opportunities***

Environmental objective	*Example*	*Business opportunity*
enhance environment	broad leaf tree planting	ethical investment
preserve the environment	ESA	subsidy for safeguarding environment
repair damage	oil slick	oil clean-up
reduce damage	lead poisoning	lead-free petrol
prevent actual threat	ozone depletion	cfc-free freezers
prevent perceived threat	skin cancer	high factor sun cream

Day (1998, Internet) points out that there is an increasing recognition that eco-efficiency is not enough on its own. Business sustainability is the logical development as businesses become more familiar with the concepts and practices of environmental sustainability. He says that there are three tiers of benefits from this sustainability, which involve increases in both potential benefits and potential risks.

1 Process efficiency – low risk and high yield benefit. This is often a simple adjustment to practices such as switching to energy-efficient light bulbs, but for a large company this can give substantial savings in energy costs.
2 Product enhancement – this involves greater investment and bigger risks, but much greater benefits to the environment and to the company. An example is the Evergreen carpet leasing system of US carpet manufacturer Interface. This involves customers leasing their floor covering carpet tiles which are selectively replaced as they become worn and the old tiles are then recycled to produce new ones, thus saving raw materials, energy and labour.
3 Market positioning and development – a company takes the initiative and anticipates trends in consumer demand and legislation to become the market leader. This allows the company to make changes at its own pace, to get ahead of competitors and to make a case for its own practices to form the basis of legislation. An example of this is BMW in Germany who initiated an innovative system of reclaiming obsolete cars, disassembling them, reclaiming useful parts and materials for reuse and creating minimal waste. This helped it to gain a market advantage through favourable publicity as well as eco-efficiency and

influenced the introduction of statutory requirements based on its own best practice. There is a growing feeling that successful companies in the twenty-first century will be those who seize environmental opportunities (Day, 1998; von Weizacker *et al.*, 1998; Welford, 1997).

Constraints to seizing opportunities

Why might a business *not* seize an opportunity that might seem attractive at first sight? Figure 5.2 shows the reasons for foregoing an opportunity. They might include any combination of the following:

1 Economic:

  (a) too expensive in relation to expected return;
  (b) demand for product or service too small.

2 Social:

  (a) involves a socially unacceptable idea;
  (b) they put off investors;
  (c) they have a negative impact on image or sales of other products or services;
  (d) not sufficiently interesting to consumers.

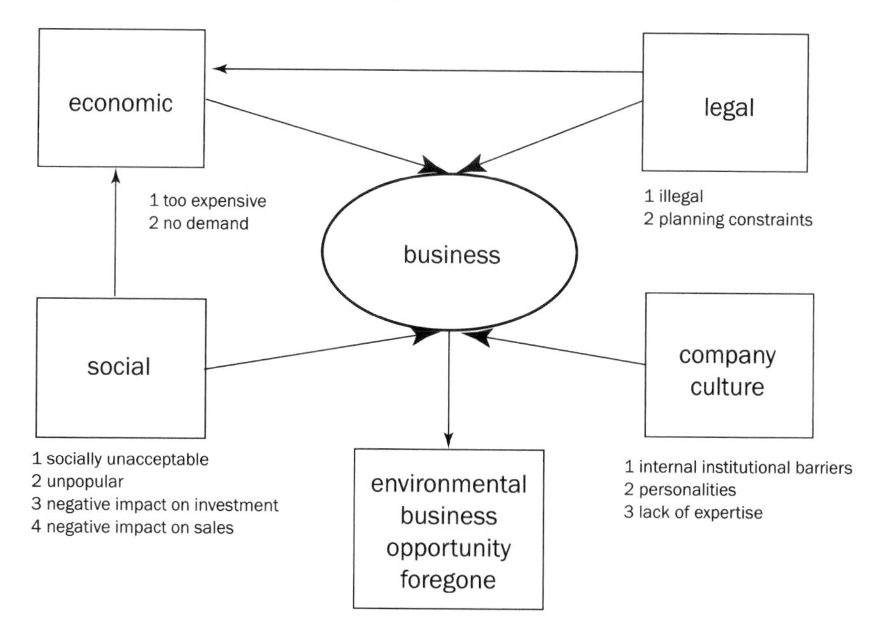

Figure 5.2 ***Constraints to business seizing opportunities***

3 Legal
   (a) contravenes existing law;
   (b) not worth waiting for the arrival of new laws or the effort of promoting new laws;
   (c) contrary to planning regulations.

4 Company factors
   (a) internal institutional barriers, interdepartmental rivalry;
   (b) personalities – 'stick in the mud' individuals resisting change;
   (c) lack of expertise to be successful;
   (d) inertia.

Physical opportunities

The physical environment offers the basic raw materials for many environmental business opportunities, such as a snow-covered mountain for a ski resort. It may be possible to develop and improve upon an environment to make it more attractive and generate a greater income than it did before. The physical environment includes not just the natural environment, but also that modified by human activity, e.g. the hedgerows of the English countryside are a physical biological element of the environment, but they were planted by people, or the built environment which may be of architectural value.

Aspects of the physical environment, and some of the potential activities associated with them, can be considered under a number of categories.

Resources

Woods, water, minerals, soils and climate form the basis of primary economic activities from forestry to farming. They also have a strong influence on tertiary businesses such as tourism, which often depends on the natural or cultural landscape for its existence. The nature and quality of the environment forms the basis of the offering. What has been created for one purpose such as farmland for food production is increasingly seen as offering opportunities for other business ventures such as farm-based recreation. Farmland offers a variety of possible recreation activities, horse riding, bird watching, painting, and walking, as well as observation and participation in country crafts, and farm activities, which many urban dwellers find fascinating as a contrast to their normal daily lives.

Natural habitats

These influence the nature and distribution of wildlife and hence the location of nature reserves and national parks. These, in turn, influence the opportunities for activities such as bird watching, hunting, or fishing.

Landscapes

Landscapes are the visible outcomes of the interaction between nature and human activities giving the distinctive appearance characteristic of a particular region which has become the primary reason for its attractiveness to visitors, who seek out viewpoints and indulge in walking, orienteering, photography, painting or simply admiring the view. Film location agencies (see Box 5.2) frequently need to find particular types of landscape or buildings for a variety of films, TV, advertising shoots etc. The landscape is their raw material.

Heritage

Ancient monuments, country houses, villages, structures (e.g. bridges) have given rise to tourism based on site seeing. Heritage centres, based on archaeological or historic sites (e.g. battlefields), industrial archaeology or country houses capitalize on the contrast between the past and present and the desire of people for 'edutainment' (knowledge-based recreation). Certain regions rich in historic remains, such as the valleys of the Rivers Nile, Loire, Dordogne or Wye, or the Weald have capitalized on their heritage and developed numerous tourist-based businesses.

Other factors

Pollution and waste give rise to damage repair and control businesses. These have a clear basis in the need to deal with pollution and waste. Environment change is a more recently recognized phenomenon that can provide businesses with opportunities, for example, ozone depletion has

Box 5.2

English Heritage and the movies

Period films and television adaptations of classic novels often need authentic locations. It is virtually obligatory for any English murder mystery to be set in a country house. Advertisers, too, frequently use historic settings for their campaigns. There is then a commercial demand for such property. English Heritage, the quango responsible for more than 400 historic properties earned £100,000 in 1998 directly in location fees, but more importantly, the films provided valuable publicity and moviegoers then visit the property. The film *Mrs Brown* (1997) about Queen Victoria and John Brown was largely filmed at Osborne House, Isle of Wight. There was a dramatic increase in visits after the film was released with £1.1 million revenue up by 20 per cent from the previous year. The effect has continued and in 1998 visitor figure were up 300 per cent. English Heritage was no doubt pleased that the film received even wider exposure on TV at Christmas 1998. English Heritage charges up to £500 a day for external shooting and up to £3,000 per day for interior filming.

Recent films shot at English Heritage properties include:

Mrs Brown (Osborne House); *Robin Hood, Prince of Thieves* (Old Wardour Castle); *Hamlet* (Mel Gibson version – Dover Castle, Kenneth Brannagh version – Castle Howard (not English Heritage); *Elizabeth* (Warkworth and Aydon Castles); *Basil* (Marble Hill House); *Jane Eyre* (Rycote Chapel) and *Sharpe's Regiment* (Tilbury Fort).

Source: M. Kennedy, 1998, "Films bring tourist boom for stately homes", *Guardian*, February, 1998, p. 8 and English Heritage *Visitor Guide* (1998).

given rise to concerns about the incidence of skin cancer and has thus boosted the sales of high factor sun cream. Worries about water pollution have led to an increase in the consumption of bottled water.

The key element in all these businesses is that without the environment there would be no business. This may seem a statement of the obvious, but the point is that the environmental aspect used in the business, be it a natural ecosystem, a landscape, or a man-made structure is as much a raw material as iron and coal are for a steelworks. Without care and protection many of these environmental resources would disappear. Using them as economic resources helps to ensure their continued existence. The non-profit organizations that have adopted business methods do so in order to generate finance to help them achieve their primary conservation objectives (see Chapter 9).

Legislative and planning opportunities

Modern business has to take account of a whole host of legal requirements ranging from health and safety to tax demands. Increasingly, environmental regulations in planning law are becoming part of the framework within which business has to operate. While legal and planning regulations can be seen as a constraint (see Chapter 4), it is also possible for businesses to use changed circumstances, created by the law, as an opportunity to stimulate the company's activities, for example, by providing opportunity for a new product or service.

There is a complex hierarchy of regulation from global to local that can affect business activities. Taking government to include everything from the United Nations to local borough councils, we can see regulations may originate at the highest levels (perhaps a UN conference or convention) but be put into practice at the local.

An example of this is Local Agenda 21. This seeks to encourage local communities to adopt integrated sustainable practices across the entire spectrum of housing, employment, transport, energy, waste disposal, recycling through to land use planning. The concept originated at the 1992 Rio Conference (the Earth Summit) and was taken up by numerous municipalities all over at the world. It is not a mandatory measure, but a voluntary one which has captured the imagination of many local communities, as in Manchester (Smith *et al.*, 1998).

International law

There have been many international conferences, conventions, protocols and treaties, which have either obliged governments to pass laws, have generated publicity to stimulate change or have created new environmental organizations. These have worked to create changes in the individual's and government's attitudes towards the environment. Business cannot ignore the dimension of international environment law for the following reasons:

- Some law is truly global and virtually every nation subscribes to it, for example, the Law of the Sea stipulating oil pollution limits.
- Many nations have adopted strict environmental standards and multinational companies need to be aware of their obligations. This has often been beneficial for countries that do not have such high

standards since a company which is asked to meet the standards in, say, The Netherlands, finds it easier to apply their unified policies in all the countries in which they operate.

- Business may regard laws as impositions or they may seek to turn them to their advantage. An example of this is the Montreal Protocol 1987, seeking a ban on CFC chemicals damaging the ozone layer. By 1992 this had been strengthened by the Vienna Convention on the protection of the ozone layer. CFCs used in aerosols, refrigeration and plastic foam were phased out as an environmentally damaging chemical. Companies like ICI and Du Pont turned this to their advantage by developing non-CFC propellant gases (Eden, 1996, Howes *et al.*, 1997). At the EU level, there may be specific European requirements with which business must comply. Directives may be issued by the European Commission, whose Member States are obliged to implement in their own legislation, for example, to establish nitrate sensitive areas, wildlife habitat protection and recycling goals.

National laws

National laws have the most immediate influence, especially for those businesses that have no overseas interests. A good example of how a company can turn legislation to its own advantage by seizing an opportunity before anyone else has noticed it, is that of Envirologic (Box 5.3). Here the company took little-known competition provisions of the Water Industry Act 1991 and created for itself a niche that promised profitable business and environmental benefits.

A company may also decide that its profitability, or indeed survival, depends on using government regulation as a basis to promote the environment. An example here is of game hunting companies in East Africa that are licensed to hunt in particular areas (Box 5.4).

Sometimes government does not enact laws to achieve its objective, but seeks to persuade the business to behave in a certain way through publicity and consultation. An example here is the UK Government's promotion of sustainable business through a consultation document on the Internet that was designed not only to provide information about sustainability but also to encourage businesses to give feedback.

Box 5.3

Envirologic – using legislation to create opportunities

The 1991 Water Industry Act resulted in the creation of privatized water companies which had in effect geographical monopolies on selling water and sewerage services in their areas. Large water users like steel works, power stations, oil refineries and breweries found themselves paying high prices for drinking quality water for low-grade uses like cooling and cleaning. Envirologic, a small company set up by a former Thames Water manager, has exploited a clause in the 1991 Act which allows for competition within an existing water company's territory. This is known as an Inset Appointment which allows a company to become the water and sewerage services provider to a large user. It buys in water and sewerage services from the water company and then sells it on to the user. It makes its profits by instituting efficiency methods and recycling. Envirologic has become the provider to Shotton Paper Mill by buying in water and sewerage services from Welsh Water which was the former supplier. It has instituted schemes to drive down the cost and shares the saving with Shotton. Eventually it plans to drill its own boreholes to access its own water and cut costs even more and sell on surplus to Welsh Water. Envirologic has similar schemes lined up with breweries, power stations, oil refineries and other large users. It is a good example of where a company has created an opportunity for itself by careful scrutiny of the legislation and evaluation of the market niche. Without the 1991 Act it would not have been able to innovate in this way.

Source: Dunn (1998).

Box 5.4

Big game hunting and conservation

It might at first seem strange that the most effective way so far devised of saving the large animals of the East African savannah is to hunt them. Big game hunting generates far more income than photo safaris. In 1997 big game hunters generated £26,000 per person while photo safaris generated only £450 per person. The pressure on habitats and wildlife arise from poverty and consequent pressure on the land for cultivation, grazing and poaching. In Tanzania the government licenses hunting and touring companies and a quarter of the fees are supposed to go to local authorities, but little of this filters down to local people who see little value in respecting wildlife. An American hunt company, Tanzania Game Tracker Services (TGTS) became convinced that poaching and land degradation were a threat to its future livelihood – if the big game disappears, so will TGTS. It therefore set up the Friedkin Conservation Fund which gives a 15–20 per cent surcharge on game fees to local communities and in addition pays rewards to people turning in poaching equipment. By getting local people to benefit from the presence of wildlife TGTS hopes to ensure the continuation of its business as well as conservation.

Source: Gough (1998).

According to Department of Environment, Transport and the Regions, sustainable business offers both opportunities and a challenge:

> **Opportunities**
> Creating new markets and developing new products
> Increasing competitiveness
> Building consumer trust
>
> **Challenges**
> Meeting increasing demand while consuming fewer resources
> Using innovation to create higher quality processes or products with less environmental impact
> Meeting the aspirations of employees and stake holders and securing the licence to operate.
>
> (DETR, 1998, p. 1)

Sustainable businesses require firms to adopt a fivefold framework of action (Box 5.5). At present this represents a campaign of persuasion that sustainability is in the long-term interest of the businesses themselves, rather than an attempt to enact legislation to force such action. It is an interesting example of the use of government resources to attempt a redirection of business ethics.

Box 5.5

DETR and sustainable business

DETR's framework for sustainable business

Actions for business

1. More emphasis on services rather than production, so as to meet people's need but use less natural resources.
2. Promote sustainable communities by contributing to the needs of employees, local community and stakeholders.
3. Avoid harmful impacts, adopt best practice, innovate to develop new products and processes which use fewer resources.
4. Reflect environmental costs in price consumer pays. Make sure the public knows why.
5. Apply effective environmental and social standards worldwide to meet international obligations.

Source: DETR (1998).

Planning laws

Planning is about the regulation of space and as such is important in its effects on determining where and how business may operate. It may create windows of opportunity by allowing or encouraging activities in designated areas, for example, a pedestrianization scheme will change the environment for shopping. Covent Garden in London is a classic case of the planned transformation of a wholesale fruit and vegetable market into a pedestrianized retail and tourist area. Many specialist retailers such as theme bookshops, exotic food and antique shops as well as the host of pavement cafés and street entertainers have established themselves in the central square. Zoning (such as green belts, national parks, conservation areas) all create conditions which may be conducive to certain kinds of business. The establishment of a long-distance footpath such as the Pennine Way encouraged the Countryside Commission to develop a bunk house barn scheme whereby farmers along the route could receive grants for renovating redundant stone barns as cheap overnight accommodation for walkers. The farmers gained an additional income, walkers cheap and convenient accommodation and the landscape retained a valuable element of the agricultural heritage.

Case study: mushroom picking in Washington State

In many parts of the world harvesting mushrooms, berries and nuts is a widespread recreational activity, usually regarded as free goods. However, increasing regulation of forest areas, either for conservation or timber production, has seen more restrictions placed on the areas in which mushroom pickers can forage. In addition, in some places there has been increasing commercial picking by individuals who may have been displaced from employment in the timber industry. A study of the Olympic Peninsula, Washington State by the Man and Biosphere Programme (Leigel *et al.*, 1998a) illustrates the extent to which the forest has provided business opportunities, which have been circumscribed by regulations to protect the environment.

The Pacific North-West forests have seen a decline in employment and incomes from the timber industry over the last two decades and as a substitute activity there has been an increase in the harvesting of non-timber forest products for recreation, subsistence and commercial reasons. Harvesting of greenery for Christmas decorations employs

10,000 people in the $47 million harvest. In 1992 $20 million was paid to harvesters of mushrooms alone. The total annual value of the mushroom industry in the Pacific North-West is estimated at over $41 million. Most of the Olympic Peninsula is in Federal or State ownership and there are increasing restrictions on where and in what quantities pickers may harvest. Traditionally anyone could pick what they wanted, but regulatory policies to conserve the forest habitat have criminalized previously legitimate practices and have limited picking to specified areas. Table 5.3 shows how different areas have different policies towards mushroom picking. The National Park forbids commercial picking and other areas have a combination of harvest limits, permits, licence fees and restricted access. Harvesting has become more difficult as there has been a policy in the Federal Wilderness Zones of blocking roads to prevent vehicular access. Nevertheless there has been increasing demand for mushrooms, not least due to the changing ethnic composition of the region with more Asians (Cambodian, Koreans, Vietnamese) and Latinos living there linked to bigger demands from ethnic restaurants. Opportunities for cashing in on a natural harvest are becoming more evident to more people, but this has resulted in both the public and private landowners introducing restrictions. Zoning the area has affected business activity (Leigel *et al.*, 1998b)

Table 5.3 ***Restrictions on mushroom pickers in the Olympic Peninsula, Washington State***

Land owners	*Recreational pickers*	*Commercial pickers*
Olympic National Park	limit – 0.95 lbs/person/day $2–3 day use fee at some car parks	not allowed
Olympic National Forest	limits – 3.8 lb/person/day $2–3 day use fee at some car parks	allowed if licensed $50 – for 14 days $80 – for 30 days $120 – for one-year
Washington State Department of Natural Resources	allowed	allowed 11.4 lb limit – 1 species 34.2 lb limit – three species
private timber companies	some limited access 5–10 % of land only	permits variable conditions

Source: Leigel *et al.*, 1998b.

Social opportunities

Business sells products or services only if there is demand for them and this is just as true of environmentally based products and services as for any other. Consumers may demand environmental goods or services because of the following:

- ethical reasons – despite the extra cost, they feel they need to buy it in order to make their contribution to the environment;
- health reasons – an environmentally sound product or service provides assumed tangible health benefits, for example, organic foods;
- economic reasons – solar heating or recycling may be seen as a cheaper way of achieving the consumers' objectives. Some green investments have outperformed others during the 1990s;
- fashion reasons – it is the *in thing* to ride a bike, use rechargeable batteries or go on eco-holidays;
- family reasons – they may feel it desirable for their children to go on wildlife safaris or drink organic orange juice.

Attitudes to the environment have changed considerably over the last thirty years and many ideas that were considered cranky in the 1960s are now widespread, for example, drinking bottled spring water or eating organic foods. As the DOE says (Brown, 1992, p. 225): 'People's attitudes to the importance of the environment relative to other major issues is strongly influenced by media coverage.' There are indications that environmental concern has become more firmly established among the population. This can be seen from numerous surveys which have been conducted (Brown, 1992; Worcester, 1997).

Green consumers

Is there such a creature as the green consumer? Most people at some time or other would buy some sort of green product or service, but does that mean they are green consumers? A thing will be purchased for its utility to the purchaser, if it happens to be green that is a bonus. Eden (1996) is fairly sceptical about the idea of the green consumer. Elkington and Hailes (1998), on the other hand, believe that almost everyone has the potential to be a green consumer. It is generally believed that the better-off upper and middle-income groups are more likely to be able to afford the premium prices for 'green' products. This does not mean that low-income families will not be green consumers. Worcester (1997, p. 165) shows that there is a 'core group of older working class people whose values are deeply green and strongly held', estimated to be about 15 per cent of the working class. What are green consumers likely to do? The DOE survey (Brown, 1992) identified various personal actions to improve the environment (see Table 5.4)

Table 5.4 ***Personal behaviour of consumers in relation to the environment***

Percentage	*Behaviour*
50% or more	picking up litter using ozone-friendly aerosols
40% or more	avoid garden pesticides
20% or more	bottle banks use reduce use of electricity take old newspapers for recycling use alternative transport to car use recycled paper make compost of kitchen waste use unleaded petrol
less than 20%	buy phosphate-free washing powder give money to green causes buy magazines concerned with environment belong to green organizations

Source: Brown, 1992, Figures 16.7 and 16.8.

Although these activities seem quite widespread and to be on the increase according to Worcester (ibid.), it should be noted that most of them are not particularly taxing of people's efforts or pockets. Few involve purchases of specific products and the product range is limited. Partly this is due to the range of questions; there is nothing about lower energy appliances, eco-tourism or organic food purchases, all of which involve substantially more commitment and expense than picking up litter, using less electricity or foregoing pesticides (of which, the last two actually save money).

Organic food is a good example of truly green purchasing behaviour. If it is to be of any value to the consumer it must be regularly purchased. Very few British farmers are organic, less than 1 per cent, and the majority of organic food sold is imported. Some imported food is organic by default rather than choice, the small banana growers of Africa do not use pesticides and fertilizers because they cannot afford them, rather than from having green consciences. However, a series of health scares, BSE, salmonella, listeria, and E-coli in the 1980s and 1990s have turned many consumers towards organic produce. Organic products are sold through:

- supermarkets (towns and edge of towns);
- farmer's markets (in towns);
- farm gate sales (at farms);
- box delivery systems (home delivery);
- local independent shops (neighbourhoods).

All but the supermarkets have fairly limited penetration according to the *Shop Around Guide* (*Guardian*, 14 November 1998, p. 15). Waitrose, Safeway, Tesco, and Sainsbury's are the leading supermarket providers of organic produce.

The technical fix route to opportunities

The DETR consultation paper on sustainable business makes much of the need for innovation in products and services. Von Weizacker *et al.*'s *Factor Four* report lists fifty innovations which it claims will advance sustainability. Welford (1997), Eden (1996) and Howes *et al.* (1997) all make the point that business tends to favour eco-efficiency as a strategy since it means less of a departure from traditional business methods. All businesses like the idea of saving costs and anything that appears to do this is welcomed. So much the better if it boosts green credentials and gives favourable publicity. The technical fix is the method of seeking a technological solution to the effects of a problem without dealing with the root cause (Figure 5.3). A technical fix may indeed solve a problem, but may give rise to new ones. More likely it will only partially solve an environmental problem. Fundamental reform is necessary to address the cause of a problem, but this may be unacceptable to the firm because it is prohibitively expensive, risky or puts the whole business in danger of bankruptcy.

As an example, consider the environmental problems caused by the car. As car use increases, there are increased traffic jams, more air pollution, accidents, more energy use, more carbon emissions and dust and more

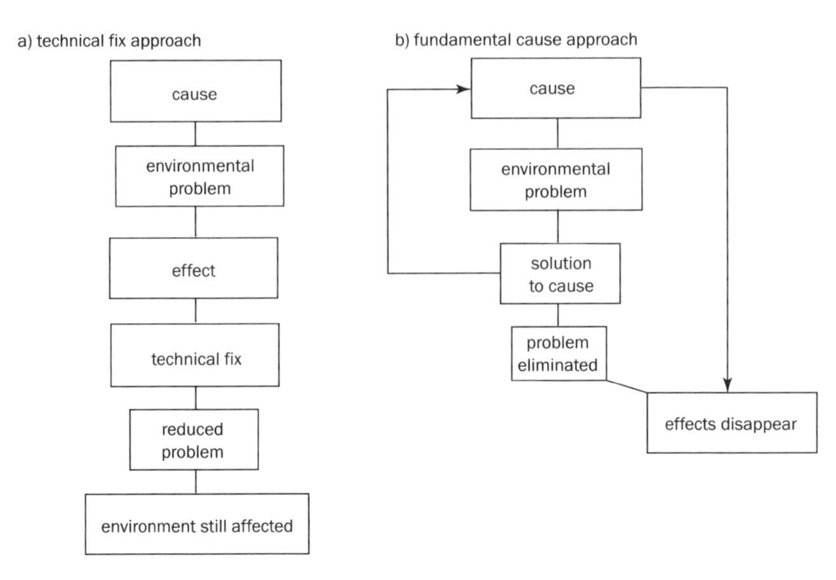

Figure 5.3 ***The contrast between the technical fix and fixing the causes***

global warming. The traditional response has been to build more roads to relieve congestion that simply leads to increased traffic and a bigger problem. However, the 'great car economy' as Margaret Thatcher once described it, is a very significant element of business. Governments fear to tamper with the car because of its economic importance and also for fear of antagonizing car driving voters. A fundamental solution to the problem of car pollution would be to actively reduce the number of cars on the road by measures including increasing the cost of motoring by road pricing, higher fuel taxes, extra licences, higher parking fees, harder MOTs, harder driving tests, etc. At the same time public transport would be made cheaper, more reliable, more comfortable, faster and safer. Thus people would still achieve their mobility objectives, but the environment would be spared considerable pollution. However, wholesale adoption of this programme is unlikely because both consumers and business would suffer in the short term. Therefore business prefers to go for technological fixes including low emission cars, greater fuel efficiency, unleaded petrol and grudgingly experiments with parking charges, taxation and road pricing. The sample could be replicated for a whole host of other industries for the principle is the same, business prefers the technical fix for its own convenience.

Summary

Business is becoming more pro-active in seizing opportunities presented by the environment and by the growth of environmentalism. Opportunities are not seized by all businesses and there are variations in the degree to which businesses see potential opportunities and have the ability to seize them. There are a number of constraints to businesses taking environmental opportunities and these need to be identified in order to overcome them. Environmental opportunities come in a number of classes including physical, legislative and social. Perhaps the biggest potential opportunity comes from the growth of the green consumer who actively seeks an environmental component to purchases. Identifying and satisfying the wants and needs of the green market could become highly significant to business in the future. Business activity inevitably affects the environment, but it need not be an adverse impact. Businesses are more likely to be keen on participating in environmentally pro-active ways if they can see benefit for themselves.

Further reading

The idea of business being pro-active towards the environment is a relatively new one. M. Smith, J. Whitelegg and N. Williams, *Greening the Built Environment* (London: Earthscan, 1998) examine the practicalities of turning a city's economy towards sustainability and emphasize the benefits of pro-activity. Perhaps the most optimistic environmental book of recent years is E. von Weizsacker, A. Lovins and L. Lovins, *Factor Four: Doubling Wealth – Halving Resource Use* (London: Earthscan, 1998), which exudes enthusiasm and positive ideas from every page. A book that is more pessimistic about what is happening now, but full of optimism about what could be done in the future to create genuine green businesses is R. Welford, *Hijacking Environmentalism: Corporate Responses to Sustainable Development* (London: Earthscan, 1997). P. Hawken, *The Ecology of Commerce* (London: Weidenfeld and Nicolson, 1993) is the book that inspired the CEO of Interface to convert his company towards sustainability. J. Elkington and J. Hailes, *Manual 2000: Life Choices for the Future you Want* (London: Hodder and Stoughton, 1998) is a highly readable book and a useful source of addresses and websites. Department of the Environment Transport and the Regions (DETR) (1998) *Sustainable Business: Consultation Paper*, Online. Available http://www.environment.detr.gov.uk/sustainable/business/consult/opps.htm, is a government consultation paper which contains examples of businesses which have become pro-active.

Discussion points

1. Why is it difficult to be sure that 'green consumers' exist? How might they be identified?

2. In what ways does the regulation of mushroom picking in Washington State provide an equitable balance between the needs of people and the environment? What would be the consequences of either banning picking completely or allowing total freedom to pick?

Exercises

1 Evaluate your green credentials

Compare your purchasing behaviour with Table 5.4. How green are you? What makes you purchase green products? Alternatively, if you are not a green consumer – why don't you buy green products? In a seminar you could compare the behaviour of different members of the group. A follow-up could be to devise a questionnaire to survey other students in the university or to do a pavement survey of the general public.

2 Creating opportunities from legislation

The case of Envirologic showed how a company could use legislation to its own advantage to create opportunities. Examine any piece of environmental legislation to identify potential business opportunities.

6 Primary industries: using resources directly

- **Relationship between resources, environment and business**
- **Primary industry and resource management**
- **Environmental impacts of primary industry**
- **Agriculture – what is it?**
- **Costs assets and externalities**
- **Farming as a business and as a way of life**
- **Environmental concern about farming**
- **Agri-environmental schemes**
- **The pressures on farming**
- **Opportunities for farming**
- **Sustainable farming**

> Although it may give the illusion of productivity, in the long run, modern farming is a failure – not only ecologically, but also economically and socially.
>
> (Goldsmith *et al.*, 1990)

> Those using the surface of the land must accept that their care must extend beyond narrow food producing objectives.
>
> (Slee, 1989)

A primary industry is one that utilizes a natural resource directly as the main purpose of its activities. It may harvest directly from the wild (rubber tapping, deep sea fishing, forestry), cultivate crops (agriculture), raise domesticated animals (ranching), extract minerals or fuels from the earth (gravel quarrying, oil drilling) or tap a naturally renewable source of energy from the environment (wind power). The key distinguishing feature which sets it apart from other industries is that the primary business purpose is to obtain raw material which is then either consumed without further processing, or is sent on to the next step in a chain of production (manufacturing) or distribution. Primary industries involve businesses that have the closest relationship to the environment, since their very existence is bound up with material resources.

Figure 6.1 shows three alternative views of the relationship between humans and the environment. In Figure 6.1 (a), we have a simple deterministic model that shows the environment as the prime mover deterministically influencing human behaviour. This model says that it is the environment that *conditions* human activity. This is a highly simplistic approach that nevertheless has a core of truth. One cannot mine coal where it does not exist, so that in a sense the geological occurrence does determine the location of the mine. However, people are not ants, instinctively responding to external stimuli without thought. Humans have the ability to respond to their environment and to consciously alter it (Figure 6.1b). Feedback mechanisms change the state of the environment and result in a different set of environmental conditions that in turn affect human behaviour. For example, depletion of fish stocks by trawling fleets may result in a severe shortage of certain kinds of fish. This leads to imposition of quotas to conserve stocks and a consequent change in fishermen's behaviour to go after different more abundant species until such time as the first species recovers. A third approach is to give primacy to the human element (Figure 6.1c). Business appraises the environment for opportunities, makes deliberate choices and through its action changes the nature of the environment, this altered environment in turn modifies human behaviour. An example might be the drainage of wet lands for agriculture changing a natural ecosystem into a cultivated one. The resulting increase in food production would allow a rise of the population that in turn would stimulate further intensification of agriculture. Each of the models tells us something about the relationship between the environment and business. Those businesses in the primary sector all have a close relationship to the resources they utilize.

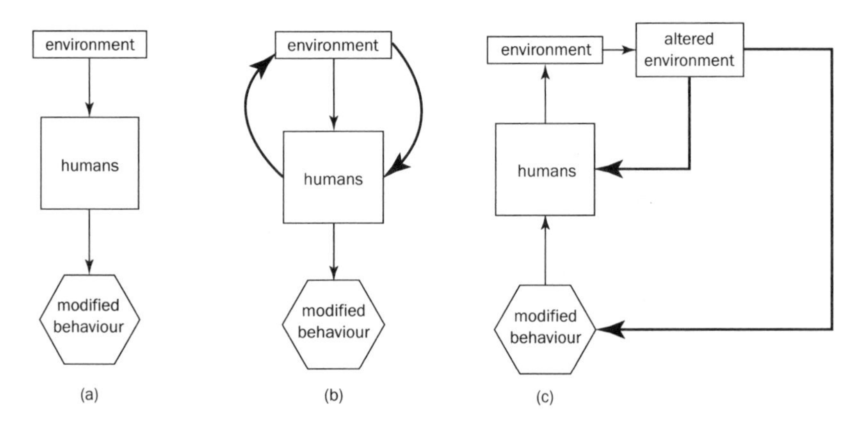

Figure 6.1 ***Models of environmental determinism***

Relationships between resources, environment and business

A resource is exploited by industry only if there is a demand for it. If there is no market where people are prepared to pay more than the cost of extraction/cultivation, then it will not be used and indeed might as well not exist. Oil was certainly in the ground and under the sea for all of human history, but it was not a resource until the late nineteenth century and not a major resource until the twentieth century. Every resource has environmental attributes (for example, forests have functions as wildlife habitats, oxygen producers, carbon sinks, watershed protectors and landscape elements as well as timber production). In some cases these environmental attributes may become more valuable to society than the primary raw material. In advanced societies, wildlife protection and landscape amenity may come to rival the primary product as a business opportunity or may act as a constraint on the primary operations, for example, logging operations in the Pacific North-West have been restricted to protect rare species of owl, and the exploitation of roadstone in the Peak District National Park of England has been subject to a variety of restrictions.

Primary production will produce waste and pollution as the inevitable outcome of operations (mine tailings, pesticide pollution, sawdust, etc.). As primary production has grown in scale, so the polluting activities have come to the attention of the public who demand reductions in pollution emissions. These demands for amelioration of waste and pollution can be seen as potential costs for businesses, who will only comply with demands if:

- they are forced to do so by regulation;
- they see the potential for cost-cutting;
- they see the potential for a commercial opportunity.

As we have seen in Chapter 5, it is possible for businesses to turn costs or externalities into opportunities. In the case of primary industries, the nature of the environment in which they operate frequently allows for exploitation of demands for the environmental attributes such as forests or farmed landscapes. Regulation in the form of government legislation is important both in setting limits on environment damage and in encouraging good practice. It may also assist in developing the use of the environmental attributes as a way of both limiting harm to the environment while protecting the industry from damage. Much agri-environmental regulation comes into this category – governments provide a wide range of schemes which are designed to protect and enhance the environment, whilst ensuring farmers do not lose income for doing so.

Primary industries and resource management

A key concept in resource management is the distinction between renewable and non-renewable resources (O'Riordan 1971; Simmons, 1996). Renewable resources are those that are capable of self-regeneration and could in theory go on indefinitely. Biological resources come into this category, plants and animals reproduce naturally and over time stocks would tend to stabilize. Humans, however, act as a super-predator and harvest these resources, thus preventing any natural expansion. Provided the rate of extraction equals the rate of renewal, the resource is in a steady state and is said to be sustainable. However, if the rate of extraction exceeds the rate of renewal, then the stock of the resource goes into decline and will eventually disappear. This is a state of unsustainability. It is possible to extinguish a renewable resource entirely by over-exploitation, for example, the rapid demise of the North American bison herds in the last quarter of the nineteenth century, or the extinction of the flightless dodo by hungry seafarers in the Indian Ocean. Although it would seem the height of foolishness for a business to destroy the very basis of its own existence, it is possible for it do so, for example, the whaling industry. Renewable resources therefore need careful management to ensure replenishment. It is possible to manage them in a sustainable way. Sustainable agriculture and sustainable forestry do exist and may be further developed under suitable circumstances (see Table 6.1).

Non-renewable resources, however, have a finite life – as the old saying goes: 'They ain't making any more'. Fossil fuels and minerals form at such infinitesimally slow rates on the human perception of time, that for all practical purposes we can regard them as a finite stock, which can be depleted or conserved but not increased. Minerals or fossil fuels are non-renewable resources and they can only be used more efficiently to prolong their lifetimes or they can be conserved by substitution by renewable resources. Fuel efficiency, for example, to extract more calorific heat per litre of oil burned is one way to extend the non-renewable resource. Even if extended, non-renewable resources will eventually be consumed. Many fossil fuels can be effectively substituted

Table 6.1 ***Primary industries and resource management***

Industry	*Resource*	*Type of resource*	*Typical management*	*Green potential*
farming	land, crops animals	renewable	some sustainable but most unsustainable	more sustainabilty required
fishing	fish	renewable	largely unsustainable much depleted	urgent need for more sustainabilty
forestry	trees	renewable some non-sustainable	some sustainable	extend sustainable to more forests
mining and quarrying	minerals	non-renewable	non-sustainable depletion	reduction recycling reuse to extend mineral use substitution
oil and gas and coal	fossil fuels	non-renewable	non-sustainable	fuel efficiency substitution
renewable energy	wind, water waves, solar	renewable	sustainable continuous	large potential

by renewable energy sources, for example, wind, water, solar, bio-gas, wood, or geothermal. These renewable energy sources have increasing attraction commercially as the technology improves and the costs come down. Wind power is a good example of a renewable energy source that is clean, freely available and dropping dramatically in price so as to compete economically with fossil fuels (British Wind Energy Association, 1998, online). As oil and gas reserves are depleted, so the laws of economics dictate their price will rise. This means they will be used for those activities which cannot easily find a substitute and which are forced to pay the higher price – so-called premium markets. If the price of oil goes up, it will be used for motor transport since there is no practicable alternative (in the short term) but not for space heating, where there are other options. As the price of oil creeps ever closer to that of the novel energy sources, it makes investment in the novel alternatives more attractive, resulting in a drop in price as economies of scale come into play. We are at the point where major oil firms are carrying out research into such energy sources, but not yet at the point of widespread adoption.

Environmental impacts of primary industry

Because of their intimate relationship with the environment, the primary industries have a widespread and significant impact .

Pollution

The non-renewable energy industries are highly polluting, e.g. oil and gas flares contribute to global warming and oil spills cause great localized harm to marine ecosystems. Mining and quarrying can also be highly polluting. Farming using industrialized methods pollutes over large areas. Pesticides, in particular, have had a pervasive and invasive influence on the environment with even quite small amounts being harmful. Fishing and forestry pollution in comparison is only slight. It should not be forgotten that these industries are part of the overall industrial process and that they both make use of manufactured products in their processes and provide the raw materials for businesses further along the production chain.

Waste

The fossil fuel and mining industries are the main culprits in waste generation. Numerous locations in Belgium, Germany, the Czech Republic, Russia, the USA, China and the UK are scarred by vast quantities of spoil. Farming produces moderate amounts of waste, mostly crop residue and animal waste, which can be recycled. However, factory farming often creates a considerable waste problem from the huge quantities of slurry which exceeds the capacity of local ecosystems. Fishing and forestry and renewable energy create almost negligible waste in comparison.

Recycling

Recycling is one of the green activities that has the potential to make significant improvements to the environment. Renewable energy is the greatest contributor to recycling, for example bio-gas. However, most renewable energy sources are minimal producers of waste or pollution in the first place. Farming and forestry have great potential for recycling.

The traditional mixed European or American farm made use of everything produced on the farm. For example, straw from crops was not waste but could be recycled as animal bedding or mixed with nutrients as feedstuff. Animal manure was used as a valuable source of soil nutrition. However, the decline of mixed farming and the rise of monoculture turned potentially recyclable materials into waste products with adverse environmental impacts since modern developed farms run on purely commercial lines tend to ignore recycling to the detriment of the environment.

Habitats

Farming and forestry have the greatest overall impact on habitats because they occupy the greatest areas of land. The type of farming or the type of forest has a profound influence on the nature of flora and fauna of a region. Changes in the system of production have dramatically affected wildlife through the shrinkage or modification of suitable habitats. Fossil fuel extraction modifies habitats since the soil, groundwater and atmosphere around oil wells and coal mines are affected by both normal operations and accidents. The *Exxon Valdez* spill off Alaska and the *Sea Empress* spill off Pembrokeshire both caused enormous damage to the local ecology, both onshore and in coastal waters. Fishing may also have a large impact on habitat through fish farming, for example, expansion of prawn farming in South East Asia has radically affected mangrove swamps.

Landscapes

Farming and forestry form important landscapes in much of the developed world. Wilderness areas are in retreat, and have to be carefully protected by national park or nature reserve status. In a country like the UK, the farmed landscape occupies three-quarters of the land and 'the countryside' is synonymous with farmed land. More intensification has led to great changes in the appearance of the landscape. Fishing only really has a landscape impact in terms of inshore fish farms. Mining, quarrying and fossil fuel extraction have a much more localized but dramatic impact on the landscape. Vast pits and spoil heaps dominate local landscapes. Offshore oil rigs may be highly visible even close to shore and in some areas oil pumps cluster in 'farms'. Renewable energy can also have a high visual impact. Wind farms sited on hills or coasts can be seen for a great distance. Whether one finds these aesthetically pleasing or monstrous intrusions is a matter of taste.

Agriculture – what is it?

For the world as a whole, agriculture, including crop land and permanent pasture, occupies 37 per cent of the globe's land area. Central America has the highest proportion at 53 per cent and Europe the lowest at 22 per cent (the European figure is heavily biased by the inclusion of the Russian Federation with only 13 per cent). The world's leading agricultural exporter, the USA, devotes 47 per cent of its land area to agriculture, while the UK with 71 per cent has one of the highest proportions given over to farming (World Resources Institute, 1998, Table 11). Farming, however, has a small share of the world's GDP, only 5 per cent in 1995. There is a clear association between poverty and dependence on agriculture. Low-income countries have an average of 25 per cent GDP from farming, middle-income countries 11 per cent and high-income countries 2 per cent. The USA, the UK and Sweden average between them 2 per cent. Only eleven countries in 1995 had more than half their GDP derived from agriculture – Burundi, Ethiopia, Equatorial Guinea, Tanzania, Uganda, Cambodia, Laos, Myanmar, Albania, Moldova, and Georgia (World Resources Institute, 1998). The proportion of rural dwellers has declined to 59 per cent in the developing world and 24 per cent in the developed world with very few of these engaged in agriculture. The OECD average is 2 per cent. Historically this is a recent and dramatic change. A century ago, agriculture was the predominant employer almost everywhere, except the UK where industrialization had come earlier. Agricultural output has expanded dramatically to keep pace with the quadrupling of population this century. The message is clear that output is up, labour is down, and that it is the industrialization of farming that has made it possible. But at what price to the environment?

Farming is distinctively different from most other business activities in that it has both a strong dependence on the environment and a strong influence upon it. For these reasons alone it deserves special attention. Oddly, most books on business and the environment ignore agriculture almost totally. Elkington and Burke (1987, p. 128) acknowledge that farmers do have an impact on the environment, but then say virtually nothing about them. Others do not even go as far as that. It may simply be the urban-industrial bias of most authors, or it may be the lingering hangover of the idea that farming is more of a way of life than a business. For most farmers in the developed world this has not been true since the Industrial Revolution and it is becoming less true even for farmers of the developing world. Pretty (1998) estimates that about one-third of the world's population is still supported by traditional farmers with the rest being supported by industrialized or green revolution farmers – in other words, commercial farmers.

What is farming? Farming is the deliberate manipulation of living organisms to produce food and other biological products such as hides, wool, fibres, oils, etc. The process of farming involves the utilization of physical and economic inputs (Figure 6.2). Some aspects of farm system are internal to the farm, for example, soils and hydrology on the physical side and tenure and fixed capital on the economic side. Others are external to the farm such as the climate and pollution on the physical side, or subsidies and labour on the economic side.

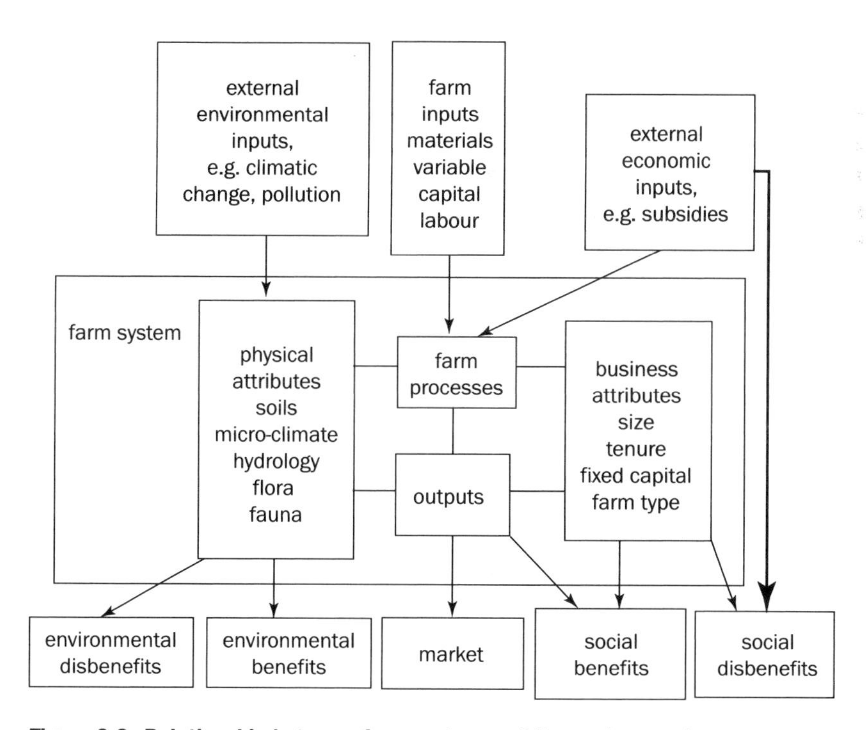

Figure 6.2 ***Relationship between farm system and the environment***

Conventionally these inputs are managed by the farmer to produce outputs that may be:

- consumed by the farm family;
- recycled to the farm system, i.e. crops fed to livestock;
- sold off farm at the market.

However, farming also produces a range of other intangible benefits and disbenefits. The environmental benefits include habitat creation, wildlife protection, landscape creation or retention and pollution absorption. Environmental disbenefits include pollution, dust, noise, habitat destruction, wildlife loss and soil erosion. In the social arena, benefits include employment, farm produce, creation and maintenance of amenity and access for recreation. These have all become increasingly important considerations in developed countries, especially where the rising costs of supporting intensive farming have come to be at odds with the desire for an accessible attractive countryside. The secondary benefits of farming are also increasingly seen as potential business opportunities for farmers to help replace more environmentally damaging activities. Social disbenefits include transfers of wealth via taxation through subsidies from urban to rural, restriction of access to intensively farmed land and loss of landscape amenity.

Briggs and Courtney (1985) have looked at temperate farming systems from a physical point of view and argue that agriculture is both a physical and an economic system. Pretty (1998) would argue that it is also a social system that underpins the whole of rural society.

The farmer has to perform the following operations:

- tillage operations – i.e. ploughing and cultivating to prepare optimum conditions for seed sowing, removal of crop residues and weeds and to improve drainage conditions;
- drainage and irrigation – to obtain the optimum moisture conditions for plant growth;
- fertilizer and manure – to provide the correct nutrient balance for crop growth;
- pest control and crop protection – to prevent pests, diseases and weeds from reducing yields and protect crop from adverse climatic conditions like wind and frost.

All these operations have to be carried out before the ultimate task of harvesting the crop.

Bayliss-Smith (1982) has examined agriculture from the ecological viewpoint and views farming as a means of transforming certain types of matter and energy into different forms of matter and energy. In this view the key variables are how much energy goes into the system and how much comes out. On this basis some traditional farming systems are surprisingly efficient. The Mareng slash and burn farmers of Papua New Guinea use no energy except human labour and produce a high ratio of energy output to inputs. Industrialized farming on Bayliss-Smith's analysis looks far less efficient with high yields only obtained by extremely high inputs of fossil fuel. A comparison of environmental impact of different beef production systems in New York State, Brazil and the Sahel (Giampieto *et al.*, 1992; see Table 6.2) showed that the more intensive systems had more deleterious impact on the environment than the less intensive. However, both studies also revealed that the yields of food are very much higher in the industrialized systems. In purely economic terms, if the environmental values are ignored, there is no contest between the industrialized and traditional systems. The industrialized systems win out every time in terms of sheer volume of output.

Costs, assets and externalities

Costs to farming are generally measured as being strictly economic inputs. In the UK, for example, there are detailed costings of inputs and outputs collected by MAFF (MAFF, 1997). Net farming income in 1990 was 21 per cent of the gross output. This varied greatly between different types of farm. Gross inputs accounted for about half of the gross output value. Fertilizers and pesticides are an important but relatively small proportion of overall costs. The biggest input is feeding stuffs – i.e. bought-in animal fodder. Much of this is imported and is a consequence of the intensive nature of UK farming that it cannot support the high livestock density from its own resources. In this sense it is unsustainable and dependent on food grown elsewhere in the world.

Farming experiences other environmental costs due to modern farming practices. In Ohio, 36 per cent of farmers experience groundwater pollution due to excess fertilizer usage (Napier and Sommers, 1994). However, most environmental costs are not paid for by farmers but by society as a whole. Pretty (1998, p. 71) estimates that in the UK alone, costs of soil erosion caused by farming practices cost between £23.8

Table 6.2 ***Comparative environmental impacts of different types of beef production systems in New York State, Brazil and the Sahel***

System	*Location*	*Former ecology*	*Fragility*	*Farm strategy*	*Ecology altered*	*Capital level*	*Only renewables?*	*Negative impacts on ecology?*
1	NY	deciduous forest	low	economic return	very high	high	no	high
2a	NY	deciduous	low	economic return	medium	very high	no	low
2b	Brazil	tropical forest	high	economic return	very high	very high	no	very high
3	NY	deciduous forest	low	economic return	medium	moderate	no	medium
4	Sahel	semi-arid desert	high	minimum risk	very low	nil	yes	very low

Notes:
Systems:
1 = beef feed lot closed system (all feed produced on farm)
2a and 2b = beef feed lot open system (some imported feed)
3 = beef feed lot organic system
4 = beef nomadic pastoralist system

Source: Adapted from Tables 2,4 and 5 in Giampieto *et al*. (1992).

million and £50.9 million annually. Pesticide removal from water costs £121 million annually and nitrate removal costs £24 million a year. In the USA externalized pesticide costs are estimated to range from $1.3 billion to $8 billion a year. It is generally acknowledged that all these costs are under-estimates and it is hard, if not impossible, to put a price on other externalities such as a loss of bird song or changes in the landscape and the demand for access.

Farm assets can be measured in the conventional way by valuing the machinery, buildings, land, animals, etc., but farming also has intangible assets such as landscape and habitats. These are notoriously difficult to put a price on, but just because they are intangible does not mean they are not assets, nor does it mean they cannot form the basis of a business, as we shall show later in the chapter. These intangible assets can be used as a basis of a business that is both profitable and environmentally beneficial.

Farming as a business and as a way of life

It is generally agreed that farming has moved from being predominantly a way of life to being predominantly a business, but that even business-oriented farmers display a wide range of motivations and attitudes (Gasson, 1973). Bowler (1985) says that the hallmarks of post-war European agriculture have been intensification, concentration and specialization. Symes and Marsden (1985) claimed that the industrialization of agriculture involves scale economies, increasing use of inputs from other sectors, resource substitution, closer integration of production and marketing and organizational features associated with industry.

There are, however, significant differences between farming and other businesses:

- There are still many farmers, so that even the largest cannot exert a monopoly compared to, say, the manufacturers of computer disk operating systems.
- Most farms are still operated by family firms, who live on the premises. Farm managers are still in the minority.
- Most farmers expect to pass on their land to a family member and therefore wish to keep the land in good heart.

- Farmers live in day-to-day proximity to the environment and therefore have considerable practical experience of environmental processes.
- Unlike, for example electronics manufacture, farming cannot start the production cycle any time. They have to wait for the appropriate season of the year to sow seeds, for example.

A well-known typology of farmers by Bell and Newby (1974) makes a useful basis for examining behavioural differences. The typology is based on two elements: market orientation and degree of manual/administrative involvement on the farm. The resulting matrix gives a fourfold classification (Figure 6.3). We have taken the classification and added comments about farming, motives, environment and subsidies. As can be seen, there is considerable difference in each of the elements between the different types. The most extreme contrast is between the hobby farmer and the agribusinessman I. A hobby farmer, with a main income derived from elsewhere, is principally interested in the amenity or environmental value of his farm. This ironically often approximates to most urban visions of what a farm should actually look like – mixed farming with low-density cherished animals grazing in small enclosed paddocks, traditional buildings, hedges and semi-natural patches of wood and water. These farmers have little or no interest in maximizing production. They are involved because they like the image of farming and it provides a bucolic retreat from their main occupation elsewhere.

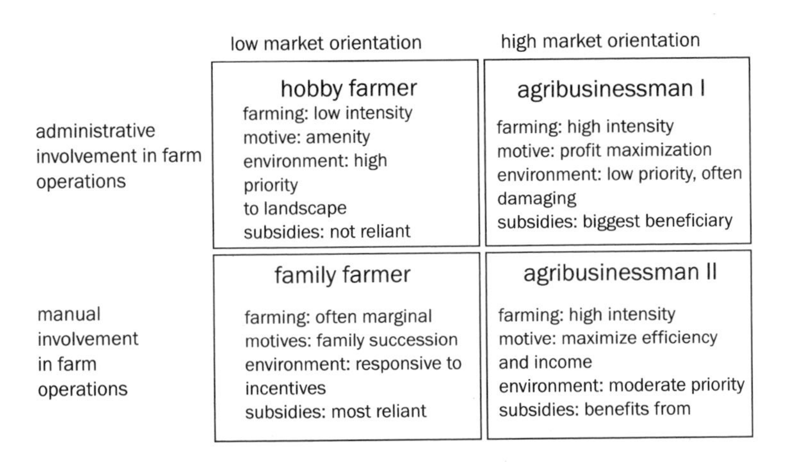

Figure 6.3 ***Behavioural differences according to farmer type***

Note: typology of farmers based on Bell and Newby (1974). Farming, motives, environment and subsidies elements added by authors.

The agribusinessman I, however, has as his main motivation profit maximization and cost-cutting. The result is high intensity farming, and high yields, but at the cost of damage to the environment. Hedgeless prairies are the most likely landscapes on this type of farm. Very often this type of farmer is a farm manager for a financial institution and as an employee he has no interest in passing on the farm to a successor. Family farmers, who are the traditional custodians of the countryside, are under threat, except for those who have the largest farms. In continental Europe family farms are frequently marginal economically, often marginal physically and are reliant for survival on government subsidy. They have the motivation to protect their assets for a successor and as Ward and Lowe (1994) have shown, farmers who plan family succession are much more likely to be interested in pollution reduction and prevention. Although under financial pressure, which could lead to environmentally damaging cost-cutting, they are more likely to be responsive to agri-environmental schemes.

Some family farms have made the transition to agribusinessman II, where although they plan to pass on the land, nevertheless they have modernized and farm to a high intensity. They wish to maximize efficiency and profit but they have stopped short of wholesale destruction of their environment. The environment is of interest to them and they will be moderately responsive to agri-environmental programmes.

Environmental concern about farming

As with many environmental issues, concern about the way farming was affecting the environment began with a landmark book by a prominent environmentalist. Rachel Carson's *Silent Spring* (1962) brought the issue of pesticide pollution to public notice by focusing on the decline in bird life. Castillon (1992, p. 202) says that *Silent Spring* 'changed the way people looked at pesticides for ever'. Ehrlich and Ehrlich (1987) also acknowledge the pivotal role of Carson's book and say that the 'widespread blatant misuse of pesticides' moved ecology 'firmly and permanently' on to the political agenda. In the UK environmental concern over the changing countryside has attracted a number of writers: Fairbrother (1970), Shoard (1982), (1987), Mabey (1980) and Harvey (1998) have all demonstrated the links between changing farm business practices and changes in the countryside.

Pesticides

Before the Second World War pesticides were highly toxic, expensive and rarely used. Wartime saw the development of synthetic pesticides, the organochlorines, of which DDT is the best known, and organophosphates pesticides, which were a by-product of chemical warfare research. DDT was widely used to kill lice and mosquitoes in war-torn areas and was hailed as a miracle pesticide from where it rapidly spread to agricultural use. However, by the late 1950s the drawbacks were becoming evident; birds were dying and there was massive decline in predatory birds in particular that suffered low birth rates due to the concentration of toxins in the food chain. By the mid-1970s over 27,000 tonnes per annum of active ingredient were being applied to British farms (Mellanby, 1981). This pales into insignificance against the 500,000 tonnes used in North America, which accounts for 30 per cent of the global pesticide market, worth 30 billion dollars a year (Pretty, 1998, p. 51). Declining wildlife has continued despite increasingly stringent rules on pesticides.

Fertilizer

Fertilizer has been a major factor in the increase of food yields worldwide. The most widely used nitrogen and phosphate fertilizers, however, do have detrimental environmental effects, notably eutrophication of watercourses, but contamination of groundwater is a growing problem and cleaning up nitrates is expensive. In 1994 world annual average fertilizer use was a 113 kilograms per hectare of crops with the heaviest applications in The Netherlands, 592 kilograms per hectare. By contrast, India used only 80 kilograms per hectare and Tanzania 10 kilograms per hectare (World Resource Institute, 1998). However, over the period 1984–90 fertilizer use in the developing world increased by 41 per cent, while use in the developed world decreased by 64 per cent. This indicates that while in the most developed countries fertilizer use is beginning to decline, more intensive methods are spreading to the Third World.

Pollution

Agriculture contributes very little to air pollution: less than 15 per cent of the UK total (DoE, 1996). Energy consumption and noise complaints are also minor, but pollution incidents and waste (about 20 per cent of the

UK total) are considerable considering the size of the sector. Slurry pollution from intensive livestock farming is a particular concern because of its geographical concentration and impact on water supplies. Of more significance is that 83 per cent of the damage caused to SSSI's (sites of special scientific interest) is attributable to farming. These impacts are in effect externalities arising from farming which are being imposed on society.

Human health concerns

There are concerns over animal health affecting human health. Intensive livestock farming is known to allow the transmission of disease more easily than extensive systems. The UK has experienced numerous food poisoning incidents in the 1980s and 1990s, e.g. salmonella, listeria and e-coli. There is also worry about the effects of hormones and antibiotics and the potential fears about GM foods. But the biggest and costliest agricultural health disaster has been BSE, which resulted in an export ban on British beef, and the expensive, £4 billion, slaughtering campaign to eradicate the disease. The human cost could be appalling because of unknown numbers of future victims of Human Variant BSE/CJD due to the long incubation period. Pesticides have also compromised human health. In the USA, 1,200 to 2,000 pesticide poisonings a year are recorded, while in the Third World an astonishing 25 million people a year are estimated to be poisoned. In the Philippines and Egypt half the farmers have reported sickness due to pesticides use (Pretty, 1998). In the UK there is continuing controversy over farmers falling ill after exposure to organochloride sheep dips.

Landscapes

Modern farming has a big impact on the landscape, mostly in field enlargement and habitat removal. The Countryside Commission study of lowland England by Westmacott and Worthington (1976), the first of a series of studies, showed the extent of change in seven lowland English parishes. In one typical parish, Leighton Bromswold, Huntingdonshire, the proportion of land under cereals had increased from 32 per cent in 1920 to 67 per cent in 1970. The average farm size increased from 122 to 317 acres, while the number of farms fell by half; the average field size had increased to 45 acres and the number of workers had declined from one to every 74 acres to one for every 240 acres. Farmers had acted for sound business reasons to eliminate landscape features. They were encouraged by the productionist ethos of MAFF which gave grants for the improvements. These have now ceased and been replaced by agri-environmental schemes.

Wildlife and habitats

Wildlife and habitats have declined as a consequence of chemical pollution and habitat change. Many once common species have suffered drastic decline. For example, the number of hares are down by 80 per cent this century, partridge by 75 per cent since 1945 (Cocker, 1998), corn bunting by 69 per cent, tree sparrow by 67 per cent, lapwing by 59 per cent, bullfinch by 58 per cent, song thrush by 55 per cent since 1970 (Fuller *et al.*, 1991). A great concern in the UK has been over the disappearance of hedgerows, which have declined from 796,000 kilometres in 1947 to 428,000 kilometres in 1990 (DOE, 1996). Hedgerows contribute both to wildlife habitat and to landscape so that they are seen as a characteristic symbol of the English countryside.

Access

Increasing demands for access by walkers and horse riders have been given the label the right to roam in the UK. This has been a political battle for many years (Hill, 1980) and many people supposed it to have been won with the 1949 National Parks and Access to the Countryside Act. However, increasing demand from recreationists has met with resistance from landowners and the right to roam became a political controversy in 1999. Even when the government announced plans to extend the right of ramblers to walk on the open land, they were watered down from earlier proposals. In many other European countries the right to roam has long been established.

Agri-environmental schemes

Governments have long controlled farm production by systems of subsidies, grants, loans tax relief and price-fixing. In the 1980s and 1990s the embarrassment of surplus production at the expense of environmental destruction encouraged developed countries to reorient goals towards the environment but emphasizing essentially the same administrative mechanism, i.e. to pay farmers to do, or not do certain things. In Germany there are large numbers of schemes in operation to encourage farmers to be more positive towards the environment. Wilson (1994) notes that schemes are characteristically large in number (91), voluntary, biased towards wet lands and meadows, often conserving very small ecosystems such as orchards and field margins, not always geographically limited (45 per cent apply to anywhere in the country), most established since 1992, have no specific time limits and have higher per hectare payments compared to other EU states.

The UK agri-environmental schemes (Table 6.3) are a mixture of geographically restricted areas (ESAs and NSAs), countrywide schemes (Countryside Stewardship, and Farm Woodland Premium), and habitat specific (Moorland scheme). The present UK agri-environmental schemes are simpler than those of the earlier decades which betrayed clear evidence of incremental additions, unclear objectives and confusing detail (Gilg, 1996). Like Germany the schemes are voluntary and even within the eligible areas uptake is not universal.

Australia's Landcare system (Box 6.1) shows how very different environmental problems result in a radically different approach compared to the European countries. Australia's farmed environment is very recent compared to Europe and there is no long heritage of settled countryside

Table 6.3 ***Uptake of agri-environmental schemes in England, 1994–98***

Scheme	*Area (ha)*	*Number of agreements*
Countryside Stewardship	106,806	6,225
Environmentally Sensitive Areas	469,121	9,201
Farm Woodland Premium	36,850	5,915
Nitrate Sensitive Areas	19,611	359
Organic Aid	4,673	101
Habitat Scheme	6,874	412
Countryside Access Scheme	1,627	126

Source: MAFF 1999 Environmental Schemes. Online. Available http://www.maff.gov.uk/environ/envsch. The figures were assembled from pages on each of the schemes.

Notes:

1 Moorland Scheme is not measured in areal terms but involves payment for removal of ewes from eligible areas to reduce overgrazing. The total removed was 3,900 from 13 agreements.

2 It is meaningless to total either the areas or agreements since farms may subscribe to several at the same time.

Box 6.1

Australia's Landcare scheme

Landcare is a government-sponsored system of community based schemes for dealing with environmental problems and protecting natural resources. It primarily involves farmers and pastoralists working in small local groups with expert advisers. There are 4,250 Landcare groups and approximately one-third of Australia's farmers belong to a Landcare group. The scheme is administered jointly by the Ministry of Agriculture Fisheries and Food Australia (AFFA) with the Ministry of the Environment. Funding via the national Heritage Trust in 1995–6 was A$ 117 million. AFFA states that 'Achieving sustainable agriculture is a major environmental goal for Australia' (AFFA, 1999). It is clear from the breakdown of the projects that the main objective is building networks of co-operating farmers who work together to improve their local environment. The main emphasis is on protecting and enhancing the environment *for agriculture*.

Example projects:

1 Farm Management 500, Southern New South Wales. This involves 500 farming families working with 15 advisers to develop computer packages with farm databases and GIS to fine-tune farming practices for specific localities. Their objective is develop sustainable technology to enhance profitability. They see the route to this being very detailed scientific measurement of things like soil variability and computer specification of precise inputs.
2 Cundumbl Landcare group, New South Wales. The area suffers from increasing salinity and bush fires. Ten farmer members of the local Bush Fire brigade form the Landcare group and from their mutual interest in reducing salinity on their properties broadened their concerns to other environmental aspects such as tree planting. The most significant feature has been the growth of co-operative activities.
3 Centralia Land Management Association, Northern Territory and northern parts of South Australia. Covering over 600,000 square kilometres, this area of arid pastoralism, centred on Alice Springs, is extremely thinly populated. Just eighty families manage an area bigger than Great Britain. The chief environmental problems relate to the aridity of the region and its associated blights of weeds, rabbits and erosion. The Landcare scheme here aims to combat these problems.
4 Woady Yaloak, Victoria. This is a mixed farming area producing crops, wool and livestock along the Woady Yaloak River. Over 2,000 ha have been affected by erosion and there are 80 kilometres of gullying. Six Landcare groups covering 220 farmers organize to combat these problems. The approach is said to be to: 'help farmers improve their profitability so they can afford to care for their environment'.

Source: AFFA (1999) online.

to be preserved. In contrast, there is the problem of an often harsh arid environment where the problem is battling against drought, bush fires, salinity and rabbits. Another contrast is that the Australian land care system is farmer-oriented. The state provides the framework, the experts and the finance but the emphasis is on the bottom-up approach with strong inputs of local knowledge and organization. Farmers are engaged by the scheme because its emphasis is not so much on conserving the environment for its own sake, but conserving it for increasing the profitability of farm businesses.

The pressures on farming

Farming experiences a number of external pressures. To complicate matters, most of these pressures are interlinked, for example, climatic change is linked to pollution but is also linked to economic, social and political pressures.

Pollution pressures

Pollution pressures arise as a result of day-to-day farm operation (Table 6.4). Many of these activities have become pollution problems because of the scale on which they are carried out. Pigs have always produced manure, but the development of large pig units containing thousands of animals and the accommodation of units in restricted areas, for example, Humberside in the UK, has meant that pollution impacts have become larger and more concentrated (Symes and Marsden, 1985). Pollution pressures become significant to farmers when society insists on the industry cleaning up its act.

Global warming

Global warming is a controversial issue which has causes mostly external to agriculture, though farming does contribute its share of greenhouse gases through methane and carbon dioxide production. The pressure on farming is more from the predicted consequences of global warming such as flooding of low lying land and changing climatic conditions.

Table 6.4 ***Pollution pressures on farming***

Farm activities	*Environmental effects*	*Action taken*
Fertilizer application	nitrate contamination eutrophication	fertilizer restrictions, monitoring, bans
pesticides	death of beneficial insects vertebrate poisonings human poisoning	licensing of products licensing of operators investigations
sheep dips	human health impacts	licensing, supervised disposal
tillage	soil erosion and consequent flooding	remedial clear-up
deep ploughing	tree root severance	voluntary restraint
increased stock density	overgrazing, erosion	stocking limits, stock removal
intensive livestock	animal health and food chain contamination	slaughtering, export bans stricter hygiene rules, inspection, herd certification, animal passports
field enlargement	hedgerow removal habitat loss, landscape loss	tree and hedge planting schemes
slurry from intensive animal husbandry	water and odour pollution	planning permissions refused; fines, codes of practice
drainage	loss of wetlands	environmental schemes
irrigation	water abstraction, low rivers	licensing limitations
waste straw burning	air pollution, fire risk	ban on burning, recycling schemes

Land use change

Farming experiences pressures for conversion of land to other uses, principally urban uses. This has been a concern for much of the century and is increasing with the expansion of space-demanding activities such as out-of-town retailing. But there is also a demand for more housing as family structures change, with more single people and the population demanding more spacious accommodation. The growth of car transport leads to greater space demands for roads and parking spaces. Pressures for land use conversion are most acute at the urban fringe (Blair, 1981a; Munton, 1983).

Social pressures

Since the Industrial Revolution most people have become steadily more divorced from the realities of rural life. Paradoxically, as real contact has diminished, so the dream of the rural idyll has increased. In a reversal of the trends of the last century and a half, the last quarter of the twentieth century has seen the phenomenon of counter-urbanization: a movement back to the countryside by urban people, but this is not a movement to become farmers, but simply to live in the countryside, sometimes to work but most often to commute or retire. Robinson (1990) shows that this is a common phenomenon in all developed countries. The new country dweller has a preconceived image of the countryside, which often does not match reality. The media, especially nostalgic period television film dramas, create images of an idealized landscape. The new country dweller does not have sympathy for intensive livestock, pesticides, herbicides, fungicides, or hedgerow removal. Even those who live in towns have strong rural yearnings. A large proportion of the population of the UK goes out into the countryside on a sunny summer weekend. There is therefore a strong social pressure on farming to conserve the rural environment in the image of that portrayed in the popular media.

Economic pressures

Farming has had to cope with a number of economic pressures which have tended to favour increases in farm size, mechanization, chemicalization, monocultures and business orientation. Among such pressures are:

- Falling world prices for most food commodities. The real price of wheat had fallen to 53 per cent of its 1960 value by 1995, that of maize to 49.5 per cent (World Resource Institute, 1998, Table 6.3).
- Increasing dependence on subsidy. CAP for EU farmers has been a powerful force towards encouraging modernization and especially increases in monocultural arable production. Ironically for a system ostensibly designed to help small farmers, it is agribusinesses which

have benefited most. CAP's Agenda 2000 seeks to reduce expenditure substantially by reducing prices, while still supporting farmers' incomes, maintaining sufficient production encouraging diversification and safeguarding the environment (European Parliament, 1999).

- Over-capitalization in machinery and buildings, partly encouraged by subsidy and guaranteed prices, has led to farmers investing hugely in tractors. In the UK there were 500,000 tractors on 280,000 farms. This works out at 13 ha for every tractor and compares with 14 ha for France, 9.5 ha for Germany, 4.6 ha for The Netherlands, and 2.3 ha for Japan. The more extensive farming systems of Australia and the USA have 149. 9 ha and 39. 5 ha for each tractor, respectively.
- Debt. Expansion during the good times, when prices were high and governments were generous and underwrote almost anything, led to many farmers borrowing to finance farm purchases, machinery and buildings. Changing prices, rising interest rates, worsening trade situations and retrenchment by governments have proved the undoing of many smaller farmers in particular. Average external liabilities for farmers (including bank loans) were £74,200 for England, £42,100 for Wales, £59,000 for Scotland and £16,500 for Northern Ireland in 1996–7 (Carrol, 1998).

Political pressures

The 'Farm Lobby' has undoubtedly lost political influence over time (see Chapter 2) but it still retains influence well beyond its electoral strength. In the EU farmers' protests often bring traffic to a halt, in the USA senators from farming states can bring influence to bear on Congress to pass legislation favourable to their farming constituents. During 1997, the UK countryside became a more prominent political issue. This was largely a culmination of a series of unrelated issues which all seemed to come to a head at once, coinciding with the election of the new Labour Government. The continuing damage to the livestock industry by the BSE beef export ban was compounded by the direct action protests on veal calf transport. Outrage at the government's proposals for massive house building programmes in the greenbelts and a private member's bill to outlaw hunting with hounds, led the Countryside Alliance, largely a pro-hunting pressure group (Hanbury-Tennison, 1997) to mobilize rural protest. As a response, the government announced the creation of the Countryside Agency to coordinate the work of the Countryside Commission and the Rural Development Agency. However, farming organizations still felt that their case had not received enough publicity and in May 1998 the Rural Charter was launched by twenty-five organizations led by the NFU and the CLA (*Farmers Weekly*, 1998a). It included as signatories, farm machinery dealers, agrochemical firms, livestock hauliers and veterinary associations as well as farming bodies. The point was made that farming is the lynchpin of the rural economy.

Opportunities for farming

A central thesis of this book is that the environment provides many opportunities for business. Farming has been adept at using technology over the last century to overcome an environmental problem and this was initially hailed as great triumph to increase output. However, farm technology has created many new environmental problems. There are hopes that the latest technology will prove more environmentally friendly (see Box 6.2). For example, *precision farming* is a package which involves detailed mapping of fields to reveal micro-spatial variations in nutrient status so machinery can apply the optimum quantity of fertilizer in exactly the right locations. Precision farming is claimed to be highly effective in cutting costs, cutting pollution and still maintaining or even increasing yields (*Farmers Weekly*, March 1999c).

Farming the subsidies

Agri-environmental schemes have become more widespread and many different schemes, e.g. access schemes, woodland planting, and countryside stewardship can all apply to the same farm. There is therefore potential to replace environmentally damaging practices without loss of income. However, as Table 6.3 shows, the extent of uptake of some schemes has been quite limited in the UK. Even in ESAs the uptake was only 40 per cent of the eligible area, which in itself is only a very small fraction of the whole land surface. It is clear that although the remuneration may seem quite generous, not that many farmers have taken it up overall. This is in striking contrast to Australia which claims one-third of all farmers enrolled in a Landcare scheme. There is also a big gap between enrolment in the two biggest schemes, the Countryside Stewardship scheme and ESAs, and all the other schemes. Set-aside has

Box 6.2

Late twentieth-century agricultural technology

Conventional technology – widespread from 1960–2000

- horticulture – automated glasshouses;
- intensive arable – mechanized operations – tractors, cultivators, seed sowing, harvesters;
- chemical farming – herbicides, pesticides, fungicides, inorganic fertilizers;
- automated spray irrigation, aerial crop operations (crop dusting);
- intensive livestock – battery farming/factory farming;
- zero grazing beef lots;
  - controlled grazing;
  - rotary milking parlours.

New technology: developing from 1980–2000

- Computerization
  - farm management, farm offices, farm records;
  - micro-climate monitoring, soil analysis;
  - temperature-/humidity-controlled crop storage on farm;
  - automated livestock housing real-time computer control.
- Precision farming
  - ICM-Integrated Crop Management;
  - no plough cultivation, direct drilling;
  - global positioning systems-micro-site mapping/precise dose application.

Twenty-first century technology? 1990s' innovations limited as yet but soon to spread fast?

- Biotechnology
  - genetically modified seeds;
  - genetically modified animals;
  - cloning;
  - growth hormones;
  - hydroponics;
  - robotic farming – robotic machinery, computer-controlled.

proved more popular, accounting for 13.9 per cent of all cropland in 1995, just slightly less than the total area of all agri-environmental schemes put together. Set-aside areas fluctuate as does the payment for it, but as agreement to set-aside land has been tied in with eligibility to arable area payments, farmers have a double incentive to apply. If they don't agree to set-aside, they don't get the other subsidies. Also, set-aside is temporary (five years), and actually involves doing more or less nothing with the land, unlike the other agri-environmental schemes which are usually longer term and require considerable work, for example, tree planting or hedge laying.

Pluriactivity

'Pluriactivity is defined as the participation by any member of farm household in income earning activities that contribute to the viability of the household' (Bateman and Ray, 1994, p. 2). Pluriactivity means farm families trying to earn money from a variety of activities that most often use a resource of the farm or skills of the farm family, but different from the main farm enterprise. A dairy farm which offers farm accommodation and sells dairy products on the farm shop is a pluriactive farm, a cottage just offering bed and breakfast without a farm is not. The idea of pluriactivity is to diversify farmers' incomes to make up for any losses in the main farm business and to wean them off production subsidies. The range of possible pluriactivities is shown in Figure 6.4. As far as the environment is concerned, many of these pluriactivities can be environmentally beneficial, for example, organic farming, agro-forestry, farm tourism. They may provide mutually reinforcing motives for adopting an agri-environmental scheme, for example, hedge planting and pond creation would help make the farm more scenically attractive and an access scheme to allow walking and picknicking would draw more people to the farm shop. Bateman and Ray (ibid.) say 93 per cent of farm families in upland Wales are pluriactive which supports the view that such activities are most likely in areas where farms are small and under severe economic pressure.

Farm tourism is an activity that has gained prominence in recent years as a potentially beneficial activity for farmers and the rural environment since it is regarded as suitable because it assists in achieving the goal of increasing farm income, retaining population but also protecting the environment (Box 6.3) (Slee, 1989; Pearce, 1990; Blair, 1987).

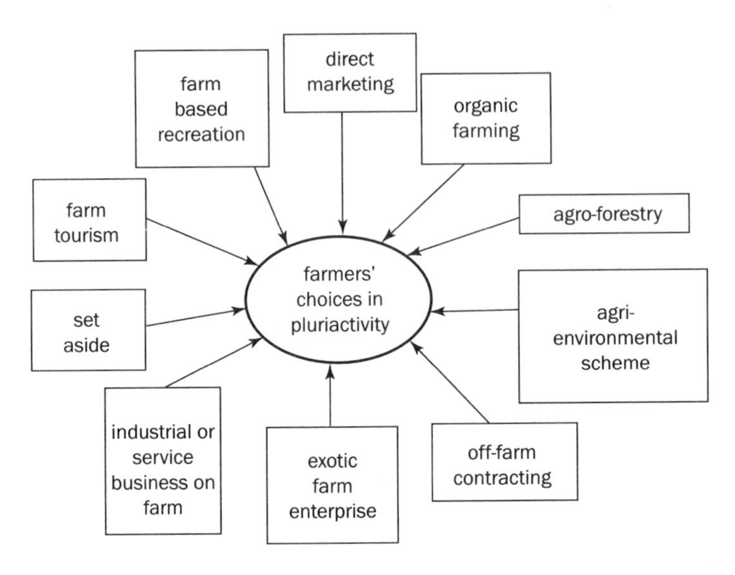

Figure 6.4 ***Farmers' choices in pluriactivity***

Box 6.3

Farm Holiday Bureau (FHB)

> Visitors are welcome to help feed the animals and join in the farming activities.
>
> (*Stay on a Farm*, FHB, 1991)

The FHB is the best-known organization which collates and publishes advertisements for farm holidays in the UK. Its annual *Stay on a Farm* guide is widely available. FHB is organized into 91 local groups (FHB, 1998), covering around a 1,000 members (numbers fluctuate annually). Although the purpose of the guide is to sell holiday accommodation, it provides a wealth of information about farm types, and farm activities for visitors. The groups are mostly biased towards the traditional holiday areas such as the South West and the North West with relatively few in Eastern England or Northern Ireland. Just over half the entries are for working farms with an average size in 1991 of 116 ha. Some 15 per cent made a special point of emphasizing the farm attractions – such as riding, fishing, farm trails, farm museum, or opportunities for visitors helping with farm tasks. Livestock and dairy farms are much more likely to be involved than arable farms, but this partly reflects the western and northern bias. FHB has counterparts in Germany, Belgium, Finland, France, Hungary, Iceland, Ireland, Italy, Luxembourg, Portugal, Romania, Slovakia, Spain, Sweden, and Switzerland.

Source: FHB (1991, 1998).

Sustainable farming

Although most regard organic farming and sustainable farming as the same thing, Pretty (1998, p. 91) makes the point that many traditional labour-intensive farm systems such as the high mountain pastures of Southern Europe are, in effect, sustainable farms. Seymour and Giradet (1987, p. 42) define organic farming as:

> A sustainable biological approach to farming, one which does not treat land like an expendable commodity. There is little erosion on an organic farm. The reason that there is little erosion is that the organic farmer operates a 'closed' farming system. That means not buying in great quantities of chemicals and not relying on the constant use of machinery, but making steady use of the natural resources of the land and recycling all its nutrients.

To be classified as an organic farm it has to be certified by an official body and pass certain stringent checks to be sure it does not use artificial fertilizers or pesticides. It is the process of conversion from conventional to organic production which has tended to retard the growth of organic farming, since for a transition period (about three years) yields may decline with cessation of chemical inputs before natural fertility can be built up, but the output does not yet qualify for premium organic prices. It is this problem that government aid is designed to overcome. The UK Government Organic Aid Scheme (MAFF, 1999) gives grants of between £25–£70 per ha over five years for non-Less Favoured Areas and £5–£14 per ha for Less Favoured Areas (since LFAs attract other subsidy of up to 300 ha per farm). As we saw in Chapter 5, demand from retailers for organic food is growing rapidly. According to Clarke (1999) all organic food accounts for 2 per cent of the EU food market and is the fastest expanding sector of food retailing with an annual growth of 20 to 40 per cent. Consumer concern about food safety has increased demand. 'Good profit opportunities exist for those who are prepared to convert. The consumers are generally ready to pay 20 to 30 per cent price premium' (ibid., p. 23). Organic produce was valued at £150 million in 1997 and £225 million in 1998 in the UK.

If there is such demand and potential business and environmental gains, why do not more farms take it up? Numbers of organic farms vary considerably in Europe with a total of nearly 50,000 farms covering 1.2 million ha, the largest proportion being in Austria. The UK has only 820 farmers and under 50,000 ha (Pretty, 1998, p. 95). This works out at only 0.29 per cent of farms and 0.26 per cent of farmed area, which is hardly a significant quantity in either economic or environmental terms. But there is increasing optimism among the farming community for organic farming as an editorial in *Farmers Weekly* (1999b, p. 5) put it: 'Organic farming with realism could become a godsend for industry. Organic production could provide a lifeline to some of the UK's most vulnerable farms particularly those in the hills. They are ideally placed to convert their less intensive systems to organic farms using family labour to replace bought-in inputs.'

An ADAS adviser in the same issue said: 'Increasingly agricultural policy has been directed towards agri-environmentalist systems. There is no doubt that the government is changing its long-term approach to farming and organic production fulfils many of its aims.'

If the area undergoing conversion is included, the UK has an estimated 200,000 ha or roughly 1 per cent under organic systems (*Farmers Weekly*, 1999a). There are a number of myths about organic farming, the most persistent of which is that it is a low productivity activity and low return business. Close financial monitoring of performance of organic farms in the USA, Europe and the UK reveals this to be untrue. The gross income of sustainable farms in the US Midwest was lower than conventional farms, but their net income was $5/ha higher since their input costs were considerably lower (Pretty, 1998, p. 117). A long-term study by ADAS in Norfolk demonstrated gross margins for an organic arable farm to be £588 per hectare higher or 45 per cent more than the conventional farm. 'The main thrust of this project is to look at the financial performance. We never expected it when we started in 1992 that the organic margins have turned out be better than the conventional' (*Farmers Weekly*, 1999a).

Pretty (1998) argues that sustainability is not something that needs overnight conversions. It can be achieved in small steps and even if a farmer only takes the first, that makes some environmental benefits, and is more likely to lead to further developments along these lines as experience is gained. Organic farming seems to offer one of the few growth areas for farmers in an industry which otherwise seems to be fairly depressed. It certainly makes immensely good environmental sense and it seems increasingly to make good business sense.

Summary

Primary industry has the most immediate relationship with the environment since it uses resources directly and therefore has a vital interest in their continued existence. However, many primary industries have not always acted in environmentally friendly ways and are responsible for a good deal of waste and pollution. Agriculture has been the main example in this chapter and we have tried to show that the environment affects farming and how it in turn has an impact on the environment. The widespread concern by environmentalists over agriculture's impacts on the environment has led to numerous regulations. The changing emphasis away from food production as the sole function of farming has led to diversification into a range of other business activities which depend on the conservation of the environment such as farm based tourism. A host of agri-environmental schemes have been initiated to encourage 'green' farming, together with the technical fix of precision farming, but ultimately the true solution lies in increasing sustainability. The increasing demands for organic produce and the repeated health scares associated with conventional farming reinforce the trends towards sustainable farming business.

Further reading

General considerations of the development of primary resource management are to be found in I.G. Simmons, *Changing the Face of the Earth: Culture, Environment, History* (Oxford: Blackwell, 1989) and for a North American perspective A. Castillon, *Conservation of Natural Resources: A Resource Management Approach* (Dubuque: Wm C. Brown, 1992) is a useful text, especially helpful for those unfamiliar with conservation terms. In this chapter we have focused on agriculture but the principles could be applied to other primary industries as well. A useful consideration of the environmental aspects of a renewable energy industry is to found at the British Wind Energy Association (1998) available online. http;//www. bwea. com

For agriculture, J. Pretty *The Living Land: Agriculture, Food and Community Regeneration in Rural Europe* (London: Earthscan, 1998) is an excellent recent review of both sustainable and unsustainable farming. For those who wish to learn more about the physical aspects of the agricultural environment D. Briggs and F. Courtney, *Agriculture and Environment* (Harlow: Longman, 1985) is highly recommended. The somewhat confusing and labyrinthine world of agri-environmental schemes is elegantly laid out at the MAFF website: MAFF (1999) *Environmental Schemes*, available online. http://www. maff. gov. uk/environ

The Australian counterpart to MAFF is: Agriculture, Fisheries and Forestry Australia whose *National Landcare Programme* (Canberra: AFFA, 1999;

available online. Http://www. affa. gov. au/agfor/landcare/nlp. html) is also easy to navigate and very informative. The UK journal, *Farmers Weekly* (widely available) is a good example of a trade magazine which puts the viewpoint of its specialist readers firmly at the forefront. A. Gilg, *Countryside Planning* (second edition, London: Routledge, 1996) has some very useful things to say about farming for production and for conservation, and M.J. Healy and B.W. Ilbery (eds), *The Industrialization of the Countryside* (Norwich: Geobooks, 1985) gives a number of detailed case studies of industrialized farming and its problems.

Discussion points

1 What are the reasons for the different rates of take-up by farmers of agri-environmental schemes in the UK and Australia?

2 If there is a growing demand for premium priced organic food and conventional farming is experiencing severe economic difficulties, why do farmers seem reluctant to adopt organic farming practices?

3 Distinguish between the terms precision farming, sustainable farming and organic farming. What are their similarities and differences? What have governments done to encourage their adoption?

Exercises

1 Examine the web sites of MAFF in the UK and AFFA in Australia. Find details of their respective agri-environmental programmes.

2 Draw up a matrix showing the comparative environmental impacts, pressures and opportunities for mining, forestry and farming.

3 Imagine you are a family farmer in the Lake District National Park. You are faced with declining prices for your main enterprises: sheep and cattle. The farm is 360 acres in size with 60 acres of valley land and 300 acres of rough hill grazing. A river runs through the valley with a large National Trust wood adjacent to it. Several footpaths cross the farm. A main road runs past the farm entrance leading to a town five miles away. There is an eighteenth-century stone farmhouse and three empty farm labourers' cottages; there are also two large barns filled with old farm machinery and tools. There are three other adults in your family and two children. The farm is in an ESA and has an SSSI (a 3-acre wood supporting diverse bird life). Consider what possible pluriactivity plan you might devise for your farm. Hint: see Figures 6.3, 6.4, Box 6.3 and Table 6.3).

7 Secondary industries: adding value and carrying the burden

- What are the secondary industries?
- Bringing the resources together
- Manufacturing processes
- Environmental implications of products after manufacture
- Business practices
- Some characteristics of the construction industry
- The construction cycle

> The key conceptual shift manufacturers need to make in becoming more sustainable is to see themselves as selling services rather than goods.
>
> (Roodman, 1999)

The secondary industries can be thought of as the manufacturing sector, the area of the economy that many regard as 'industry'. These are the industries discussed in the majority of the literature when they talk of the environmental impacts of industry. This is not surprising since this is the sector from which much of the pollution and waste products emanate. It is also the area that many of the world's largest companies operate in and results in the production of all the world's tangible products – everything from oil tankers to microprocessors and from beer to heavy chemicals.

What are the secondary industries?

Manufacturing is the core of the secondary industries. Raw materials and components are brought together and manufactured into either an end product or a component for some other manufacturing process. In economic terms resources are transformed giving value added. So, for example, when a ton of potatoes are manufactured into a ton of potato crisps the value of the potato rises from somewhere between £70 and

£120 for the potatoes at the farm gate to about £10,000 for the crisps on the supermarket shelf. An essential feature of these industries is that some tangible product results at the end of the process.

There are a wide range of processes encompassed within the secondary industries which makes generalizations difficult. However, many of the environmental concerns surround heavy industry and those industries handling particularly hazardous, toxic or environmentally sensitive material. It is this area on which most of the literature has concentrated. Environmental groups for just these reasons have singled out the heavy chemical industry, pesticide and food processing industries for criticism. The chemical industry, oil refining and metal smelting, including the iron and steel industry, are all associated with significant amounts of waste products and pollution. All are also major users of energy. Other manufacturing industries that have close links to biological processes, such as pesticides, also affect the environment. In addition, food manufacture, the textile industry and leather tanning all have the potential to create considerable pollution of the environment. The disposal of organic wastes from these processes inevitably causes considerable disruption to ecosystems. A further industry where pollution is potentially a major problem is paper and pulp manufacture. Mills require considerable quantities of clean water but often result in the widespread pollution of rivers downstream. The secondary industries also include less controversial areas and areas which are often considered outside manufacturing *per se*. Of these we have selected the construction industry as a case study since it brings to the fore a number of interactions with the environment.

The secondary industries use processes that are associated with three distinct stages in environmental terms. First, raw materials are selected and brought together for manufacture. Much of the concern surrounds the extraction of raw materials. Second, the actual manufacturing process has the potential to pollute and to produce waste materials. Third, there are environmental considerations of the finished products once they have been manufactured and sold. This includes the issues surrounding their disposal and possible recycling.

Bringing the resources together

Any manufacturing process requires a number of inputs. Of these, three have important environmental concerns. These are raw materials, energy and land on which the manufacturing plant is to be constructed. In addition, there are requirements for capital, labour, knowledge, a market and further additional factors which have little environmental importance and so are not considered here. Manufacturing has some influence on the exploitation of resources. Although in most cases the provision of basic raw materials falls into the primary sector, manufacturing has some say on the nature and amount of raw materials extracted. To begin with, a manufacturer has an influence from where they source their raw materials. As resources are used up and they become scarcer, there is a tendency for price to rise and for the quality of the resource to decrease. This is because the highest quality resources are the first to be consumed. Once these supplies are exhausted, new supplies are sought. However, first, these may be of lower grade and this will mean more of the material has to be processed, resulting in greater energy inputs and more material having to be extracted from the ground. Second, it could mean supplies are used which contain more impurities than in the original supplies, which has the potential to result in more pollution. Third, it might require more overburden to be removed to reveal the new supplies. Finally, new supplies may have to be sought from further afield or from environmentally sensitive areas. In all these cases the new supplies will be associated with a greater environmental impact than for the original supplies of the raw material. There are other implications of resources being worked out. As the price rises, only those products which can bear the increase in price will use the raw materials, so the material will be used only in premium markets. Only those processes for which there is no alternative, or those processes that can pass on the increased costs, will continue to use the raw materials. Where there is an alternative source of material, then these will be used, assuming that the price is lower. The alternatives may result in less, or more, environmental costs than the original source. Where materials become scarce this may be due to a natural shortage or may have been created artificially. This may result from an administrative restriction imposed by licensing agreements. Many forestry agencies now enforce a system of felling licences which imposes a non-market limitation on the exploitation of the resources and various quotas exist which limit supplies of fish going into the fish processing industries. Where the resource availability becomes particularly acute, there may be an outright prohibition on its use, and then substitute materials may need to be sought. It is possible this transfers the exploitation to a further source, although increasingly there are moves towards environmentally friendly and sustainable sources of raw materials. Governments impose various protectionist measures, which secure raw materials for a home industry but may restrict supplies for export. Governments also attempt to limit exploitation of natural resources in the hope this may increase the world price and hence the value of the commodity. Political and administrative instruments are of considerable importance in determining the availability of a whole range of materials.

Sometimes, where environmental awareness is raised, it may be that market pressures reduce the value of raw material. The high profile campaign against the slaughter of seal pups for their coats led to a decrease in the value of pelts, since the market decreased. However, such examples of environmental pressure of this kind leading to a reduction in the price of a environmentally sensitive raw material are rare. Sometimes supplies are restricted by legislation preventing exploitation and this is one of the main weapons used to preserve natural resources. Sometimes this leads to distortions in the market behaviour for such commodities. An interesting case is that of ivory where despite an international ban on the trade of ivory, prices on the black market continue to demonstrate a market requirement for one of the most environmentally sensitive of raw materials. Similar trends can be seen in the trade in various bits of tiger, rhino and other endangered species' anatomy, all presenting considerable problems for the conservation and resource management of such species.

Economic and environmental pressures in some cases have led manufacturers to look for substitute raw materials and components for their products. New timber products, mainly featuring different kinds of laminate structures, have resulted in a decline in the demand for tropical hardwoods for furniture manufacture, and the novel use of plastics of various kinds has removed much of the need for basic raw materials. However, despite the move to lighter construction using less raw material and for substitution, the demand for raw materials continues to rise. With the industrialization of an increasing number of countries there is still pressure from the manufacturing sector on raw materials of animal, vegetable and mineral origin.

In the 1970s much was made of the increased exploitation of the earth's resources (Meadows *et al.*, 1972; Ehrlich and Ehrlich, 1970). The doomsday scenarios that were forecast at that time have not occurred. This, however, is due to further resources having been discovered rather than a slowing down of the speed at which resources have been used. The pressure exerted on natural resources by manufacturing industry continues unabated.

The second characteristic of manufacturing industry is its use of energy. For some industries this is an important input and most manufacturers who have produced environmental policies try to reduce their energy requirements. For now it is worth noting that it is an important input with environmental ramifications, whether the energy is produced within the plant itself, through the burning of hydrocarbon fuels, or whether it is bought in from a utility, most often in the form of electricity.

The third environmental resource used by manufacturing is that of land. It is not simply the total quantity of land which is used by factories that is important, but rather the fact that the suitable land may have other uses. This may result in land use conflicts since industrial land may be suitable for other purposes such as recreation, agriculture or residential areas. In addition, the environmental impact of a factory may extend beyond the perimeter fence, in the form of pollution, traffic congestion or the need for waste disposal. Heavy industrial plant such as chemical complexes, steel works or oil refineries require large areas of flat land close to transport links. This often corresponds to coastal areas, in particular, estuarine locations. These areas are also often of considerable environmental importance for sea birds and as a wildlife area with recreational potential. The inevitable production of air and water pollution, the risk of oil and chemical spills and the required change in habitat bring out one of the classic land use conflicts between business and the environment which we see repeated in numerous locations. These include the mouth of the Rhône to the west of Marseilles on the fringes of the Carmargue, on Teesside, at Milford Haven on the fringes of the Pembrokeshire National Park and on the shores of Chesapeake Bay on the Eastern Seaboard of the United States. This list could be extended to the majority of estuaries in industrialized countries. Most industrial plants are located in areas where industry is already well established. Here the land has been zoned for industrial use. Many environmental conflicts, however, arise away from such areas where proposals are put forward for a plant in a possibly environmentally sensitive area not zoned for manufacturing. Companies do not choose to locate factories in environmentally sensitive areas for perverse reasons and so the conflicts which result require the resolution between the need for the plant to be placed at that particular location with the need to preserve the environmental value of the area. Sometimes the need for the plant is brought into question but more likely it is the particular location that is questioned. In such cases the Not In My Back Yard (NIMBY) syndrome is often the key factor in the development of the site with environmental protests being directed to highlight why a particular location is unsuitable for development.

Away from the traditional heavy industries new industries are now evolving which do not rely on bulky components. With lower transport costs these industries, e.g. the micro-electronics industry, are more footloose as to their location. That is, they have the ability to locate almost anywhere. In brownfield sites in traditional industrial areas, they can take advantages of various subsidies and tax breaks and available supplies of skilled labour. However, they can also locate in rural or semi-rural areas where the labour force takes the opportunity to live away from an urban area. The development of business along the M4 corridor between West London and Bristol has resulted in dramatic changes to the rural communities and small market towns of that region of England over the last twenty years. As the idea of locating small light industries in the countryside develops so there is also the potential for a whole range of environmentally and rural-based industries to blossom, as we shall see in Chapter 9.

Manufacturing processes

Manufacturing processes inevitably produce waste products and pollution. Waste is seen as part of the process, whereas pollution is seen as an inevitable consequence of the process that should not happen in the perfect industrial process but which, in practice, results in the degradation of some physical resource. This is most usually the air, water courses or the ground. Sound and visual impact may also be included under the broad banner of pollution.

Waste products may be toxic or they may be inert, but bulky and require disposal, for example, slag from steel production or fly ash from power stations. Heat may also be a waste product. For the manufacturer the creation of the heat in the first place is expensive and results in the use of valuable resources. The dispersion of large quantities of heat into the atmosphere contributes to global warming. The combined effects of numerous sources of heat in industrial areas can lead to atmospheric conditions which may contribute to pollution levels by modifying the natural atmospheric circulation or result in other localized climatic effects such as changes to rainfall regimes (Goudie, 2000). The discharge of large amounts of heat into rivers results in a decrease in dissolved oxygen levels and modifications to the metabolism of aquatic species. This is particularly important in summer when flows are lower and atmospheric temperatures higher. If the heat from industrial processes can be recovered and reused, then this will bring benefits for the manufacturer in the form of lower energy costs and also benefits for the environment.

Toxic waste products can be seen as falling into one of two categories – each with a different set of environmental factors. Waste products can be identified as either conservative or non-conservative. Conservative products do not break down readily in the environment. They may be either dissipated throughout the environment or may be concentrated in particular locations. For example, heavy metals, which find their way into water courses, tend to accumulate in estuarine mud and either never break down, or take many generations to become dissipated. The management of such wastes revolves around the total amount of waste dumped and on the effect that prolonged exposure and accumulation of such wastes have on the environment. Non-conservative wastes are capable of being broken down by natural processes. Hence, bacteria in the water will naturally break down sewage and organic wastes from food processing operations discharged into rivers. Moderate amounts of such wastes can therefore be dumped with no long-term degradation of the environmental quality of the water course. This means that there is a zero cost for the discharge of such wastes. There are, however, limits to the amounts which can be assimilated into the environment in this way. As the concentrations of such wastes increase, so there is more work for the bacteria to do. This results in an increased biochemical oxygen demand on the water, which means that less dissolved oxygen is left in the water. The dissolved oxygen content is the prime indicator of the health of the river. When dissolved oxygen levels approach zero due to the action of bacteria, further breakdown of the organic waste is not possible. The river loses its ability to sustain aquatic life in the river. Fish and plants die. There is a build-up of anaerobic compounds, this is indicated by the presence of hydrogen sulphide (bad egg gas), algal blooms develop and the river is said to be dead. Wastes often continue to be discharged into the river with no chance of being broken down. The management of non-conservative wastes in such an environment revolves around the rates at which the wastes enter the environment and the rates of breakdown under any set of environmental conditions.

Environmental implications of products after manufacture

When does a manufacturer's responsibilities end in terms of manufactured articles? Clearly a manufacturer has a responsibility to ensure their product does not fail on its first outing. Consumer legislation requires that products be fit for the purpose for which they are sold. Furthermore, a product cannot be constructed in such a way that it is unsafe or even that it results in excessive environmental damage. However, at this point we enter a grey area where manufacturers' responsibilities, either legally or morally, start to become rather less clear. Manufacturers, by and large, do not have a responsibility to take back a product for disposal once its useful life is complete. Try buying a lawn mower, using it for ten years or so and then contacting the manufacturer to tell them it has broken and they can come and collect it! However, there are examples where market pressure or company policies have resulted in just that happening. Such practices tend to be as a market requirement to ensure a new sale rather than as a result of altruistic environmental thinking. Buy a new washing machine and your old one will be removed. However, the disposal will be provided by the seller of the new machine not the manufacturer from whom you purchased the original machine.

Common sense dictates manufacturers cannot be held responsible for the environmental damage (or any other kind of damage) resulting from their products whether they were used for the purpose for which they were designed or not. A manufacturer of garden spades cannot be held responsible if an individual uses one of its spades to dig up a Site of Special Scientific Interest. The manufacturer of a box of matches, whose product is used to start a fire, which destroys an area of scientifically important heathland cannot be held responsible – even though the matches were manufactured for the purpose of setting fire to things!

However, in recent years the responsibilities of manufacturers have increased as the result of both environmental and consumer legislation. Market forces have also required manufacturers to consider the uses to which their products may be put. Even a short time ago refrigerators were disposed of on scrap heaps. Now the manufacturer of the machine is required to use refrigerants that are not excessively harmful to the ozone layer. They are also required to warn that care has to be taken in the disposal of the appliance and to ensure a mechanism exists for the disposal of the refrigerants. Increasingly where components may be recycled, this is indicated on the device. The collection and subsequent recycling of a host of components including exhausted batteries, cartridges from printers and tyres are now common place. Hewlett Packard operates a laser toner recycling system with a return label (printed in Swedish), but the scheme only operates in certain European countries (which does not include the United Kingdom). Sometimes the value of the product ensures this takes place. On other occasions, such as the disposal of old car tyres, a charge is levied on the disposer to cover the environmentally friendly disposal mechanism.

Manufacturers, through the design of products, can influence the environmental impact of products long after they have left the factory gates. This can be done through reducing pollution levels, curtailing energy usage during operation or taking into consideration what will happen to the product at the end of its productive life. A number of influences come to bear on manufacturers in how they design and market their products in the light of environmentally induced market and legislative forces. For example, car manufactures for many years have sold cars on the basis of their fuel consumption and their low polluting characteristics (as well as a series of other characteristics such as power and speed, which negate any environmental gains from other aspects of the design).

Business practices

Manufacturing industry has the ability to make changes to its products in the light of environmental influences. Not only can it change the finished product but it can also change the manufacturing process and even the components. Many manufacturing companies have environmental policies and most of these strategies are directed towards very similar ends. These are to reduce the use of resources, energy, waste and harmful pollution and to promote recycling and reuse of finished products. Ricoh's (1999) strategy is based on the three pillars of energy savings: pollution reduction, recycling and resource conservation. Many firms have now developed such strategies. Although the cynic might say any such aims are more to do with good intentions and overall statements of what might be in the future rather than actual actions, many firms have realized that such plans make good business sense. These policies contribute towards ecological modernization whether explicitly stated or not.

It is in the nature of environmental economics that many projects yield long-term returns while the financial pressures on manufacturing firms are for a short-term pay back. Hence there is a tendency that while wishing to adopt environmentally sensitive processes, the resources are never quite there to allow their adoption.

Various outside influences that come to bear on manufacturing industry, however, force them to move towards environmentally beneficial ends sooner rather than later. First, they are under pressure from regulatory authorities, such as the Environment Agency, to reduce the risk of pollution incidents and to reduce waste emissions. The BATNEEC principles require factories to be monitored frequently and to seek to reduce emissions. Various other pieces of European legislation also point in this direction. As plant becomes obsolete and as integrated pollution control becomes more and more developed, so firms must adopt cleaner technologies. This is often linked to the conditions of licensing agreements. The possibilities of prosecution and fines, or of suspension of their discharge licences, can mean bad publicity that pushes companies to comply with environmental objectives.

In the light of such pressures firms can adopt a number of different strategies. They can circumvent at least the spirit of remedial measures, or they can be more pro-active and try to make genuine reductions in the knowledge that such moves, in the long run, will bring lower costs. There is a close line between doing as little as possible and just complying with regulations and doing nothing and taking the risk of prosecution. This may occur where existing plant has only a few years to go before dramatic reinvestment will be required. One controversial approach that firms can adopt is to move its more environmentally unfriendly activities overseas to where environmental legislation is less strict. This has been a controversial area since some products that are banned in one country may be produced and sold in other countries. So, for example, DDT, although banned in Western countries, is still manufactured and finds a market in some developing countries. Manufacture is cheap and given a choice between a sub-optimal but cheap product and no affordable product the moral dilemmas are considerable.

Some characteristics of the construction industry

At first our decision to use the construction industry as an example of a secondary industry might appear rather strange. Why not consider the

Table 7.1 ***Main environmental impacts from various construction works***

Works	*Impacts*
houses, commercial and public buildings	countryside, urban environment, heritage
roads, tunnels and transport infrastructure	countryside, air pollution, noise, congestion
harbours, breakwaters	coastal tides and currents, migratory birds and marine life
dams, irrigation schemes barrages	hydrology downstream, river and ecosystems, agricultural land and fishing, settlement pattern
land reclamation works land drainage	natural wetlands, microclimate, flora and fauna

heavy chemical industry or the textile industry that are manufacturing industries that cause pollution and waste and whose operations create environmental concerns? This is exactly the reasoning why they have been used as exemplars in countless other text books. A more detailed consideration, however, shows that the main characteristics of the manufacturing sector are all present in the construction industry. Furthermore, a wider perspective of the main characteristics can be given from a slightly different viewpoint. The construction industry offers particularly good examples of where the social implications of the products are more keenly felt in environmental terms.

The construction industry is involved in the fabrication of a wide variety of different buildings and structures. Think of the construction industry and what comes immediately to mind are buildings and probably houses. However, the industry covers a wide range of structures as outlined in Table 7.1. This list will reveal why the term buildings and structures are preferred to the single word buildings. While most of this list is associated with similar environmental issues, some are concerned with very specific impacts, for example, dams have a particular impact on the aquatic ecology downstream of the dam. For most buildings and structures the environmental impacts can be analysed in terms of two principal groupings: those impacts resulting during the construction phase and those which occur during the lifetime of the structure. These are considered more fully under the construction cycle below.

Buildings as a generalization result in considerable environmental disruption and environmental impact during their construction phase. Once built, their impact is determined by the use to which they are put. So an estate of domestic houses will have smaller and different environmental impacts than a large factory. All buildings occupy land and there may be land use conflicts with adjacent buildings. The environmental impact of an intensive pig rearing plant may make it incompatible with residential housing next door. Almost all buildings make a visual impact on the landscape and much is made in planning legislation about this issue. Since buildings are usually designed to stand for a long time, any impacts they have are going to have will persist for many years. There are therefore demands that they are built to standards that ensure a long life for the building and care is now taken to ensure the energy demands are not going to be excessive over the life of the building.

The construction industry also builds roads and other elements of the transport infrastructure. The construction of roads causes disruptions and considerations have to be given to the environmental impacts of roads once built. There is a fundamental question to be asked with road programmes of not only how and where they should be built but whether they need to be built at all. The road construction programme has probably provoked more protest in the UK in recent years than any other environmental issue. *Cause célèbres* have included the schemes across Twyford Down and the Newbury by-pass. Tunnels and bridges are further elements of transport infrastructure, both of which have environmental impacts both in their construction phase and a lasting impact when in use. So the M6 not only has an impact on the area through which it passes, it also exports an impact to the Lake District, since it has opened up that area to more people and hence a greater environmental impact. Of less public note, but of considerable local environmental importance, are harbours which can be seen as part of transport infrastructure. They have similar impacts on the environment to breakwaters, barrages and other coastal defence measures. They disrupt natural currents and tides which in turn change the ecology of the area as a result. Major projects, such as the Cardiff Bay scheme, have resulted in a full environmental impact assessment, where the anticipated impacts have been appraised. Under some circumstances, where the demand for land in coastal areas is acute, land reclamation projects have been undertaken. Major schemes such as the new Hong Kong International Airport or the works along the Zuider Zee are obvious examples where there are widespread impacts on many aspects of the environment.

Dams, irrigation schemes and barrages present an unusual but important set of environmental impacts. During their construction phase watercourses are diverted, site roads built and the environment generally disrupted. After construction, land is flooded, displacing the activities that took place immediately upstream of the dam. This includes wildlife, agriculture and settlements. Downstream of such structures the natural regime of the river is modified. This has implications for fishing and agriculture since the nutrients that were traditionally relied upon to fertilize the fields no longer arrive. Sediment yields are lower downstream of the dam, resulting in erosion and channel pattern changes. The nutrients and sediment do not just disappear, they are deposited in the reservoir. The build-up of nutrients can lead to algal blooms and the deposition of large amounts of sediment. These combined effects can result in the reservoir's water capacity being dramatically reduced within a period of a few decades as silting occurs.

The scale and importance of the construction industry should not be overlooked. Smith, Whitelegg and Williams (1998) using information from Anink *et al.* (1996) point out:

> It is estimated that the building sector is responsible for 50 per cent of the material resources taken from nature, 40 per cent of the energy consumption [in Europe] (including energy in use) and over 50 per cent of the [Dutch] waste generated by society.

These large and impressive numbers result from the industry's need to extract large quantities of cheap bulky raw material: sand, aggregates, limestone (for cement), brick, earth and rock. These raw materials are cheap to extract and process in bulk. The industry is also associated with the movement of large quantities of ballast and landfill in the construction of many projects. Vast quantities of earth are moved for roads and reservoirs and even in relatively modest projects the site has to be excavated, levelled, footings dug, etc. Sometimes the whole shape of the ground is transformed. Not only does the process require considerable energy using large machines but also the natural drainage, vegetation, soil and ecology suffer considerable disruption and permanent change.

The construction industry is a good barometer of levels of economic activity. It is also a good example of the way in which environmental well-being is associated with economic well-being. During troughs in the economic cycle the construction industry is among the first to cut back, laying off labour and reducing capital expenditure. Competition between companies becomes intense for the limited number of contracts available. Money is tight, as are margins. At such times commitment to environment investments, which by their nature have relatively long pay-back times, seem less important than short-term expediency. When economic circumstances improve, the industry shows all the symptoms of a boom and there is a rapid increase in activity. More projects are undertaken and, while environmental concerns are taken into account for each individual project, the combined impact of the numerous projects, at such times, cancels out most of the environmental advantage and the total size of the ecological footprint is increased.

The construction cycle

We have already hinted that the construction industry should not be seen purely in terms of the erection of various buildings and structures. Before building ever commences, as shown in Figure 7.1, there are a number of events that will already have happened. It is at the concept stage that the overall objectives of the project are assessed. It is necessary to determine what is to be done and to ensure in overall terms that the project's objectives are achievable. At this stage some appraisal of the overall costs and benefits are determined to ensure that it is worth proceeding with the scheme. This is the stage which determines what is to be built and why. Then there is a detailed planning phase that determines how the project will be built and what are the detailed implications of how it will be built. We have already shown in Chapters 3 and 4 the way in which cost benefit analysis and environmental impact analysis in its various forms are used in this process. There is a closely monitored planning system that operates in most countries. This process ensures that construction takes on board consultations with other parties affected by the project. It also ensures the structure complies with the appropriate

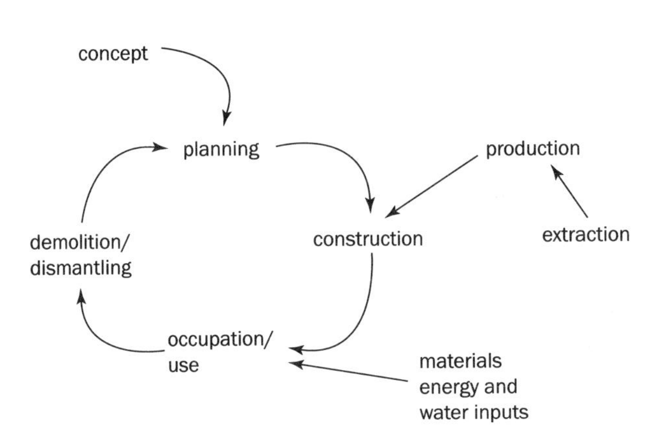

Figure 7.1 ***The construction cycle***

building regulations. In some countries this process is either weakly structured or it may be loosely or corruptly policed. In such cases there is the possibility of structures being erected which are both environmentally and structurally unsound. The zoning of land, which in England and Wales is laid out in Local Plans, ensures that in most cases incompatible buildings are not built next to each other. In National Parks, Conservation Areas and Areas of Outstanding Natural Beauty there are special safeguards to protect the environmental value of the sites.

Once the construction phase of a project has begun, there follows a period of environmental dislocation. Habitats may be destroyed. Dust, mud and disruption are inevitable. During the construction phase sediment yields of both suspended and dissolved sediments in streams increase dramatically. Since it is usual for the topsoil to be removed (often for later reinstatement) the whole biotic community is severely disrupted. It is sometimes possible to leave in place large trees and this has the effect of re-establishing some continuity soon after reinstatement; nevertheless the local physical environment of the site will be severely disrupted. Substantial volumes of raw materials must be secured and brought into the area. There is a trade-off in terms of cost, quality of materials used and the desired final look of the project. It is likely that substantial quantities of brick, cement, concrete, wood, hardcore and a variety of manufactured articles including glass, plastics, metal piping, reinforcements and steel work will be required. The manufacture of all of these has environmental implications as does the costs involved in transportation before building even begins. Often there is a push for local materials to be used to ensure buildings blend in visually with the surroundings. This has the effect of stimulating the local extractive industries, but results in environmental costs associated with quarrying and processing of the local materials.

There is much that can be done to reduce the overall environmental impact of buildings by careful selection of building materials. This requires calculations to be done in the planning stages and incorporated into the construction. There has been much work done, particularly in Scandinavia about the environmental impact of various materials through life cycle analysis (LCA). This technique examines the environmental impact of materials throughout their life from their extraction, through production, use and finally demolition. Materials should come from sustainable sources and require low inputs of energy in their manufacture. Materials such as aluminium should only be used if other advantages outweigh the energy costs of fabrication. They should have a low impact on the environment during use. The materials should have good insulation properties, so that the building will require less heating or cooling during its life. Low maintenance materials are favoured since they require less attention during their lifetime. There is a preference for materials that can be reused in their present form and to those which retain a high monetary value. Box 7.1 shows one such method of evaluating the environmental impact of various building materials. Very often there are trade-offs to be made in selecting one material as opposed to another. Softwood is a sustainable raw material but it rots easily and so its life is short. It can be preserved but only by coating it with toxic compounds or covering it with a synthetic layer. There are environmental costs associated with either action since it may make subsequent reuse difficult or impossible. Cost is always another factor to be taken into account. There is little point (from the economic point of view) in spending money on additional insulation if the pay-back times of any savings are of the order of eighty years. While much work has been done on the impacts of various materials, it is difficult to extend the techniques to whole buildings since each tends to differ from similar buildings due to different use, different microclimates, soils, humidity, etc.

When the building has been completed and handed over to the customer, the work of the construction industry is essentially finished. However, the way the building is used largely accounts for its environmental impact. Increasing use is being made of environmental audits to measure waste and energy use of existing buildings. There has been a shift in thinking over the decades in how buildings are used. Once buildings were built to last – and they did just that. Now there is increasing emphasis on the fact that they may change their use through time and so increasingly consideration is given to the idea that new uses may be found for existing buildings. This may influence their original design and construction resulting in the use of removable interior partitions, false ceilings and more versatile trunking and ducting.

Eventually buildings, or any other structure, will either wear out or the purpose for which it was constructed will disappear. This offers either an environmental opportunity or an environmental nightmare. In either case there is a likelihood that some building work – either refurbishment or demolition will be required. There are numerous examples of where old buildings or constructions have found new uses, offering a recreational, entertainment, public service or even commercial second life. Old railway tracks have been turned into cycle ways or long distance footpaths. Canals no longer required as commercial waterways have

Box 7.1

The environment preference method

The environment preference method was developed in The Netherlands in 1991 for comparing the environmental impact of various building materials and products. The various materials are compared and then ranked in terms of their environmental impact. The method evaluates the impact throughout the total life of the product. Hence it takes into consideration the environmental cost of extraction, production, the impact during use and that of disposal.

The environmental factors taken into consideration in scoring the materials are:

- scarcity of the raw materials required for the product;
- ecological impact caused by extraction;
- energy used in production and assembly including transportation costs of materials;
- water consumption;
- attendant noise pollution;
- harmful emissions including impact on ozone layer depletion;
- any contribution to global warming and acid rain;
- health risks to humans either in production or in the building;
- contribution to any physical disaster;
- repairability of product;
- potential for reusability;
- waste created.

By considering all the above, the various alternative materials are ranked on order of their environmental reference. So, for example, for the internal doors of buildings honeycomb construction with hardboard skins was the preferred option, followed by European softwood, then sustainable plywood/chipboard and least favoured tropical hardwood.

Source: Anink, Boonstra and Mak (1996).

found new life as recreational waterways. Old factories, mills and warehouses built in the nineteenth century to last for a very long time have found new uses as museums, sought-after inner city residential developments, or public buildings. Country houses, whose upkeep in the light of death duties and a changing social structure, have become museums, public buildings, commercial corporate headquarters or leisure centres, as discussed in Chapter 9. Some industrial sites have become wildlife refuges and public recreational areas. The thirst for industrial archaeology and heritage museums has given a new lease of life to ex-coal mines, slate quarries, wind and water mills, steel foundries, quarries, railway sheds, paper mills, breweries, prisons – the list is almost limitless. Such enterprises all solve what otherwise would be a difficult problem of demolition, which would have numerous, mostly detrimental, environmental impacts.

Very often the ground surrounding industrial sites has become polluted by a variety of toxic compounds. Sometimes building are demolished if they possess no architectural merit, they are unsound or the cost of refurbishment is prohibitive. Society takes an inconsistent view of derelict buildings and plant littered over the landscape. Anything recently abandoned is regarded as an eyesore and environmentally detrimental. However, mineral workings dating back to the Bronze Age within the Dartmoor National Park are regarded as of interest and contributing positively to the landscape. Demolition is costly but if re-establishment of some semblance of the former landscape can be achieved, this is regarded as a suitable alternative to doing nothing. The worst case scenario, in environmental terms, is if nothing is done with the building or construction and it has no useful purpose. The most glaring example of such a case, with long-term environmental impacts, is that of nuclear power stations. There are now an increasing number of such sites which have been decommissioned. Due to the physics of radioactive decay they will continue to present an environmental risk for many generations to come.

In recent times consideration has been given to the way in which materials from old buildings can be incorporated in new buildings. This is distinct from reusing the building. When buildings are demolished, a distinction should be drawn between high grade and low grade use of recycled materials. High grade use is when the mortar is removed from bricks and they are used again as bricks. Other examples are when slates are removed or fireplaces recovered all to be incorporated into new buildings. Often these materials carry a price premium over new materials. Low grade use is when walls are bulldozed and the brickwork crushed to be used as hardcore. Increasingly buildings are being designed so they can be dismantled rather than demolished. This practice is used in other industries, notably car manufacture, whereby components are assembled in such a way that they can be taken apart again at some later time. Components are not bonded together and composites are not used if they cannot be later separated. It is then possible to reuse the various components, for example chipboard can be reprocessed. However, if it is bonded to plastic laminate or painted then separation is not possible and recycling becomes difficult.

Summary

The chapter examines the main characteristics of the secondary industries. They bring raw materials and components together for some manufacturing process that gives value added to the components. In all cases there is a tangible product produced by the process. There are environmental impacts at all stages in the process: in bringing the raw materials together, in the assembly process, later when the products are in use and, finally, when they have to be discarded. The example of the construction industry is discussed using the concept of the construction cycle as shown in Figure 7.1 to identify the impacts at the various stages. The chapter includes some discussion of the way manufacturing businesses have adjusted their operations in the light of the environmental influences to which they have been subjected.

Further reading

Many texts on environmental management consider the effects of pollution and industrialization. D. Welford and R. Starkey (eds), *Business and the Environment* (London: Earthscan, 1996) and D. Smith (ed.), *Business and the Environment: Implications of the New Environmentalism* (London: Paul Chapman, 1993) both look how manufacturing industry has changed in the light of environmental influences. M. Smith, J. Whitelegg and N. Williams, *Greening the Built Environment* (London: Earthscan, 1998) examine particularly the built environment covering the subject far more fully than we have attempted. The concept of sustainability is one that we have only mentioned but a topic to which considerable attention has been given. A. Blowers, *Planning for a Sustainable Environment* (London: Earthscan, 1993), P. Nijkamp and A. Perrels, *Sustainable Cities in Europe* (London: Earthscan, 1994) and M.J. Breheny, 'Towards sustainable development', in A. Mannion and S. Bowlby (eds), *Environmental Issues in the Nineties* (Chichester: Wiley, 1992) all cover the issue of sustainability and the challenges that it presents.

Discussion points

1 In what ways could manufacturing businesses reduce their environmental impact?

2 Identify the nature of the construction industry's environmental impact.

Exercises

1 For any building you have access to (a college building, your own home) conduct an environmental audit, concentrating on building materials, heating, lighting and insulation. Has anything been done to reduce environmental impact? If not, what could be done?

2 For any manufactured product, draw up a flow chart showing the environmental impact at each stage from raw material to final disposal. Try to find the origin of each component part and work out how far it has travelled. Estimate the overall distances travelled for the whole product. This can be as sketchy or accurate an exercise depending on how much time and effort you are prepared to put into it. Even a rough 'thought experiment', however, will stimulate some ideas about environmental impact. In a seminar situation, different students could compare a variety of products.

8 Tertiary industries: the hidden environmental issues

- **The importance of the tertiary sector**
- **Environmental impacts of the service sector**
- **Retailing and the environment – why retailing matters**
- **Characteristics of modern retailing**
- **Retailing impact on the environment**
- **Case study: Sainsbury's environmental impact**
- **Environmental concern by retailers**
- **Environmental pressures on retailers**
- **Environmental opportunities for retailing**

> Supermarkets are popular because they are both cheap and convenient, at least for those with a car. We vote for them with our feet and wheels. But we don't always like what happens as a result.
>
> (Elkington and Hailes, 1998)

Despite the growing dominance of the tertiary industries in all parts of the world, and especially in the developed world, relatively little attention has been given to either their impact on the environment or how they utilize the environment in their business activities. This may be because:

- in comparison to agriculture or manufacturing they appear to depend far less on physical resources for their basic operation;
- their impact on the environment is less obvious because they often deal with a more intangible product. But the fact that this is less obvious does not mean that it does not exist. Compared to an oil refinery, a supermarket seems to be much less of a polluter but it is not without impact. The impacts are there but they are more subtle.
- the environmental pressures and opportunities for service businesses may also seem less obvious or pressing than those for the other sectors, but again they do exist and they do matter overall because of the size of the tertiary sector.

The aim of this chapter is to show the service businesses are significant in environmental terms.

The importance of the tertiary sector

It is likely that the majority of readers of this book will be, or have been, employed in the tertiary sector of the economy. Since at least 1980 the majority of employment in developed countries has been in the tertiary or service sector (Price and Blair, 1989). The service sector is increasingly dominant in the world economy. Of the world's total GDP of $27,680 billion in 1995, 63 per cent came from services. The largest six economies with a GDP of over $1000 billion (the USA, Japan, Germany, France, the UK, and Italy in that order) had an average of 66 per cent derived from services, with the leading nation, the USA, obtaining 72 per cent (World Resources Institute, 1998, pp. 236–7). Since these countries between them account for 65 per cent of world GDP, it is clear that services are of paramount economic significance. Even the low-income developing countries have 35 per cent of GDP attributable to services while the middle-income countries have 52 per cent (ibid.). Not only is there a close connection between levels of wealth and services at a national level, but, in general, individual service employees are higher paid than those in other sectors.

Broadly speaking, anything that is not primary or secondary falls into the service category. Although services are extremely diverse, they have one significant feature in common, they do not extract any materials from the environment as the primary sector does, nor do they manufacture anything like the secondary sector. A huge variety of classifications have been proposed for the service sector (Gershuny and Miles, 1983; Daniels, 1982; Price and Blair, 1989), but for our purposes a simple categorization distinguishing between producer services and consumer services will suffice (Table 8.1).

- *Producer services* are those whose customers are other industries, for example, an accountant may have a shopkeeper or a farmer as a client. A computer consultant may service a manufacturing firm or a theatre.
- *Consumer services* are those who serve the general public directly, for example, supermarkets, cinemas, hairdressers, or dentists. Some services may have both consumer and producer customers, for example, a bank. It is also generally agreed that this classification can be cross tabulated with public and private services.

- *Public services* are those provided by national, local or international government, for example, probation services, police, the Inland Revenue, local authority planning services and UNHCR are all public services. The general distinguishing feature is that the services are provided because society deems them necessary and in general they are not commercial profit-driven activities. However, in recent years the distinction has become less clear-cut with competitive tendering, privatization and contracting out of formerly public services to private companies, for example, private prisons.
- *Private services* are unambiguously concerned with making money and exist to satisfy demand for public wants and needs, for example, shopping, entertainment, travel, security, health, sickness, and self-improvement. Some services have always had a dual nature and exist in both public and private spheres, for example, healthcare and education, with a changing balance between the public and private. However, there are some core state services that no government, however keen on the principles of the market would privatize, for example the military services.

Table 8.1 ***A typology of service industries***

	Public services	*Private services*
producer services	civil service	accountancy
	local government	stock-broking
	waste disposal	commercial property
	air traffic control	wholesaling
	environmental health	merchant banking
consumer services	education	retailing
	health	banking
	social services	tourism
	public transport	real estate
		education
		health

Note: The examples are illustrative and not intended to be an exhaustive catalogue.

Note also that many of the 'public services' have been privatized in many countries but since they are often closely regulated by governments, they have been left in this category.

Environmental impacts of the service sector

What environmental impact does the service sector have? To judge by the amount of space in the literature, very little, though this is perhaps due to preferences and preconceptions that only the primary and secondary sectors really matter. However, we intend to show that the service sector does have an impact and that it does matter. Table 8.2 shows a comparison of relative environmental impacts for a selection of the main tertiary industries.

Areas of environmental impact are evaluated for each industry with a very broad subjective evaluation being made. The intention is to provide a broad-brush picture of the relative importance for different industries of their various environmental aspects. As a generalization, it is clear that those industries which deal with tangible physical elements have more impact than those that do not. In other words, retailing, transport, property development and tourism all utilize the physical environment directly. Retailing is about physical transfer of goods through space from one ownership to another and tourism depends on transferring people to places where they use those spaces for recreational activities. Therefore space-using activities have a bigger impact on the environment than services such as finance, telecommunications and public administration, which generally deal in intangibles. However, this is not to say that those services have no impact. Every activity that occupies a building uses energy in heating, lighting and equipment. They contribute to pollution through the travel of their employees and clients, produce waste from canteens, consume water and materials and almost certainly produce large volumes of paper waste.

Other effects of apparently low or negligible impact industries such as finance, are much more subtle. A bank may not directly pollute the environment, but may well be the source of the finance that has permitted a manufacturing firm to pollute. An insurance company itself as an office-based activity is low impact, but it can have an effect upon the environment through the operation of its premium policies. For example, private electricity generators in the USA have found it virtually impossible to insure nuclear power plant against accidents. This has been a powerful disincentive to invest in more nuclear power stations while in the UK it has had the effect of delaying or preventing the privatization of state nuclear plant. Some services are so intangible as to defy even the most ingenious efforts to detect environmental impact. An Internet service provider would seem to be the perfect example of the service business that does not have any physical product as such, existing solely in cyberspace. Does it cause much environmental impact? None? What about the millions of unsolicited Internet starter CD-ROMs pouring through letter boxes? The vast majority of these go straight into the

Table 8.2 ***Environmental impacts of service sector industries***

Industry	*Impacts*					
	Pollution	Waste	Recycling	Habitat	Landscape	Environmental concern
retailing	some	great	some	great	great	great
transport	very great	some	little	very great	very great	very great
telecoms	little	little	little	little	some	little
finance	little	some	little	little	little	little
property development	some	great	some	great	very great	very great
public administration	little	some	little	little	little	little
defence	some	great	little	some	great	great
education	little	great	some	some	great	great
health	some	great	little	some	some	great
tourism	great	great	little	great	very great	very great

Note: The impacts are shown as relative to each other in the view of the authors.

dustbin. Every service activity has some environmental impact, however slight.

Retailing and the environment – why retailing matters

Retailing is one of the major service industries, employing about ten times as many people as farming and about 11.5 per cent of all in employment in the UK (Nielsen, 1994). Other developed countries show similar proportions, Greece 9 per cent, Ireland 8.4 per cent, France 7.5 per cent, Germany 7 per cent (Eurostat, 1989). In most developed countries around 40 per cent of consumer expenditure goes on retailing (McGoldrick, 1990). Retailing is clearly an important business, but is it an important one to the environment? A number of reasons why it is of significance to the environment can be put forward:

1 It involves close regular contact with consumers, everybody is a shopper, and from this it has:
   (a) strong all-pervasive influence on consumer behaviour through stocking policy. If major retailers decide, for example, to stock only CFC-free aerosols, then this is what consumers will have to use.
   (b) Rapid response to consumer pressures. In the first three months of 1999 major retailers announced their own policies on labelling of GM foods or bans on GM ingredients in their own brands due to extensive media coverage and widespread public unease. This was despite government assurances of the safety of GM foods.
2 It has close regular contact with suppliers and is therefore able to specify exact requirements and ensure manufacturers' and producers' compliance. This is important with regard to products and their packaging but it can also extend into working practices, for example refrigeration standards.
3 Spatially widespread, but hierarchically concentrated in and around major cities, retailing mirrors the population distribution. Therefore any environmental impacts for good or ill are clearly obvious to the public.
4 It involves large amounts of transportation, for goods deliveries and consumer travel. This has implications for energy use, pollution, carbon emissions, climate change and land use conflicts.
5 It uses large amounts of packaging. Goods have become steadily more packaged, partly to ensure hygiene and long shelf life for food, partly

to prevent damage in transit, partly to prevent theft (for example, of videos), partly to protect against threats of contamination by blackmailers or terrorists (security seals), and partly because elaborate packaging provides a vehicle for advertising.

6 The food sector uses large amounts of energy for refrigeration that has the same sort of environmental impact as for transport, but there is also the issue of the ozone destruction caused by refrigerant coolant gases.

The retailing environment, see Figure 8.1, is a complex one. Retailing occupies a position in the commercial chain closest to the consumers. It receives its supplies from a number of sources either directly from producers or indirectly via wholesalers. It owes a large share of its allegiance to its shareholders but because of its dependence on consumers, it has to take account of other stakeholders. Retailing is a highly competitive business and despite concentration into fewer businesses, consumers have a choice as to where to shop. Reputation is a fragile commodity and retailers have to be alive to consumers' concerns which can arise rapidly due to pressure group or media activity.

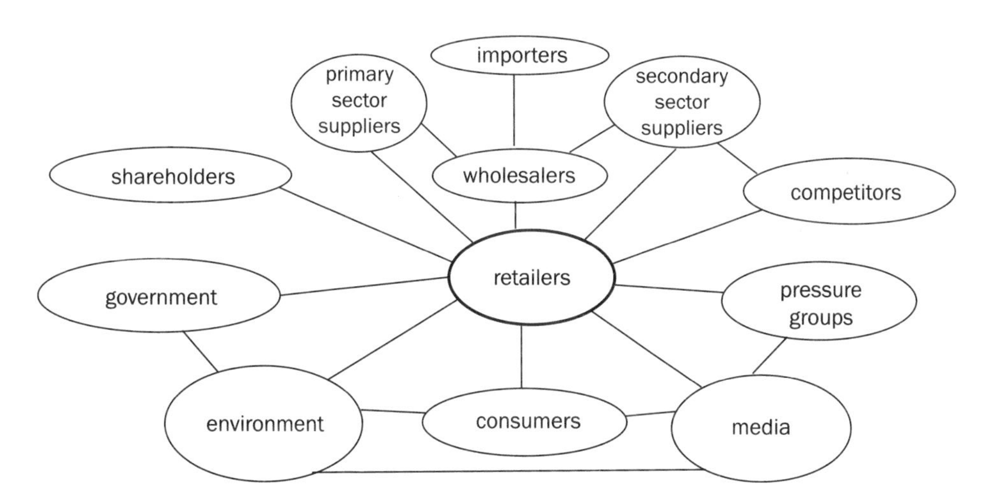

Figure 8.1 ***The retailing environment – overview***

Characteristics of modern retailing

Retailing is very diverse in format and is constantly changing. Different balances of types exist in different areas. Some, such as mobile shops and the village stores, are mostly found in isolated rural areas, while enclosed shopping centres are only found in or on the edge of larger towns. Retailing has also expanded away from the traditional shop or shopping area into venues such as transport terminals, e.g. airports or railway stations. In the past thirty years, as large firms have increasingly dominated retailing, especially in the food sector, the phenomenon of concentration has become more evident in some sectors than others. It is particularly notable in groceries, toys and DIY. This is important for the environment because large chains adopt uniform policies and therefore whatever they do, be it fascia design, car parking policies, location policies or recycling, will have a much wider impact than anything the independent store could do.

Like other business and cultural changes in the Western world, many trends in retailing began in the USA (Rogers, 1991; Carlson 1991). The impetus towards out-of-town malls patronized by car-owning middle-class consumers began in North America in the 1950s, but it spread to Europe from the 1970s onwards (Burt, 1989). Increasingly, there is internationalization of retailing with some companies becoming truly global, for example, IKEA and Toys 'R' Us (Treadgold, 1990).

It is a retail truism that there are three important things in retail: location, location and location (McGoldrick, 1990). Store location is an inherently environmentally dependent variable (in both the physical and business environment senses) and, furthermore, stores exert important environmental impacts. Table 8.3 lists the traditional and modern patterns of retail location. The traditional pattern has not been totally displaced by the modern one, but has suffered diminution, very much so in the case of the USA, to a lesser extent in the UK. Changes in locations since the 1960 have been widespread and have had a number of causes, arising from both within the industry and from changes in society at large (Box 8.1). Readers are directed to the many texts of retail change for further details

Table 8.3 ***Geographical patterns of retailing***

	Pattern of retailing
traditional	centralized
	hierarchical
	town centres (high streets)
	local neighbourhood stores
	corner stores
modern	decentralized
	out of town
	non-hierachical
	superstores
	retail parks
	retail warehouses
	regional shopping centres
	home based

Box 8.1

Retail location change and causes of change

Locational changes over the past thirty years

- out-of-town stores/edge-of-town stores proliferate
- enclosed shopping centres (in or out of town)
- retail parks on old industrial sites
- decline (death?) of the high street
- revival of the high street
- niche shopping areas, e.g. tourist areas, conservation areas
- massive decline of village stores
- rise/decline/revival of town and country planning controls on retail locations
- increased mobility/greater willingness to travel longer distances
- rise of the Internet

Causes of locational change

- spatial redistribution of the better-off population to suburbs, urban fringe and country towns
- corresponding decline in inner cities – spatial segmentation/polarization – poorer consumers stranded in inner cities
- regional change – decline of traditional manufacturing areas, growth of sunrise industrial areas, resort and retirement regions and service-based city regions
- lifestyle changes, e.g. increased home ownership – more demand for DIY
- social change – more working women, more demand for shopping in evening and weekends near to home rather than in the city centres

(Bromley and Thomas, 1993; McGoldrick, 1990; Wrigley, 1988; Jones and Simmonds, 1990).

Retail impact on the environment

Retailing has a very wide range of impacts on the environment but most retailing texts give very little consideration to the physical environment except where it impinges upon store location and then mostly as a planning constraint (McGoldrick, 1990; Jones and Simmonds, 1990). Almost nobody has taken a comprehensive view of retailing's impact on the physical environment.

Wherever they are located, stores take up space, displace other uses and have a visual impact on the landscape. A superstore with car parking can easily occupy 4 to 5 acres and even a medium-sized shopping centre can be up to 10 acres in size. A large centre can be enormous, for example, Bluewater Shopping Centre, opened on 19 March 1999 near Dartford in Kent, covers 240 acres, incorporates 320 stores, 13,000 car parking places and numerous restaurants and leisure facilities (Hetherington, 1999). Such stores have a profound environmental impact on the localities in which they are found, e.g. land and water use, pollution from traffic and energy use. They also have a serious economic effect on neighbouring retail areas, for example, Merry Hill on the outskirts of Dudley, West Midlands, has had a negative impact on Dudley town centre (Williams, 1991). In the USA the abandonment of inner-city areas because of the rival attractions of malls on the periphery by more upmarket retailers has contributed to the process of the decay. To many people the 'city' is synonymous with the main shopping area. In many European cities shopping development has been constrained by the necessity to conserve the historic heritage but in some cases the needs of building conservation has aided commercial revival, for example, London's Covent Garden.

Case study: Sainsbury's environmental impact

Sainsbury's provides useful evidence for the environmental impacts of a large store chain (Sainsbury's, 1999). Sainsbury's is one of the 'Big Four' superstore chains in the UK with 392 Sainsbury's stores, 13 Savacentres and 225 Homebase DIY/garden centres. Its environmental report shows clearly the nature of the environmental impacts and logic tells us that the other big store chains must have similar impacts. Energy consumption is the third largest cost element in running the store after employment and rent and rates. Sainsbury's group has an energy bill of over £59 million a year. Typical store energy use is shown in Table 8.5 with the largest share going on refrigeration. Refrigeration has increased in recent years due to the growing demand for frozen convenience food. More energy is also used overall because of increased opening hours and more 24-hour stores. Sainsbury's has appointed energy awareness officers and installed an energy monitoring system which has resulted in £3 million savings over three years. It is encouraging suppliers to reduce energy use and plans to increase supplies of renewable energy from the present 1 per cent to 10 per cent by 2020. At its new Greenwich Millennium store it

Table 8.4 ***Sainsbury's air emissions***

Causal activities	*Outputs*	*Effects*
goods transport	CO_2	Climate change
refrigeration	NO_x	Ozone depletion
air conditioning	SO_2	low level air pollution
own brand manufactures	Particulates	
fuel sales	VOCs	
agricultural production	methane	
customer travel		
employee travel		
business travel		

Table 8.5 ***Typical store energy use***

Activity	*% of total energy use*
refrigeration	48
heating and ventilation	18
in-store lighting	14
other lighting and electrical	12
bakery	8

plans a 50 per cent reduction in energy costs compared to a conventional new store of the same size.

The company is well aware of the causes of its air pollution and of its effects (Table 8.4). It includes as causal activities, not only its store operations but also the pollution travelling further up the supply chain in agriculture and manufacturers and by its customers' journeys to its stores. Road transport is a major source of pollution. Sainsbury group's HGV fleet and its contractors travel a total of 131.9 million kilometres a year (Table 8.6). This on its own is 0.45 per cent of all UK goods vehicle journeys. The fleet emits 136.9 million tonnes of carbon dioxide or 0.4 per cent of all UK carbon dioxide emissions due to road transport (as calculated from

Table 8.6 ***Sainsbury's road transport and the environment***

Vehicles	*HGV million km travelled 1997/8*	*CO_2 emissions 000 tonnes 1997/8*
Sainsbury's own fleet	21.1	20.2
Sainsbury's contractors	92.1	98.3
Homebase contractors	18.7	18.4
totals	131.9	136.9

Source: Sainsbury's (1999).

Sainsbury's and DOE figures). Sainsbury group produces 251,898 tonnes of waste annually, of which 57.87 per cent is recycled. The figures for just this one chain suggests that retailing overall must be highly significant as an impact on the environment, but it is impossible to put together an overall accurate picture since reporting in this depth is still rare. Sainsbury's are to be complimented first for measuring these impacts and, second, for being willing to be open about them.

As well as the items mentioned above Sainsbury's also considers its impact on water use, water discharge, land use, raw materials, nuisances and the natural environment. This example shows it is possible to give a detailed account of a retail firm's environmental impact. Simms (1992) in a review of the major grocery chains environmental policies and strategic management found although Morrison, Safeway, Tesco, ASDA, CWS and Sainsbury's all had some sort of green policy, only Safeway, Tesco and Sainsbury's really seemed to understand what greening activities actually meant. Sainsbury's green activity appeared to Simms (ibid., p. 36), as 'intelligent and co-ordinated'.

Environmental concern by retailers

Retailing depends on selling products to consumers and is therefore an active promoter of the consumer society. As such, it has often attracted criticism from environmentalists. It does appear from the Sainsbury evidence that retailing has a significant impact but the same evidence suggests it is possible to mitigate quite a lot of the impact by careful management. This is an eco-efficiency approach which aims to incorporate the environment as good business practice, reduce costs, contribute to the environmental improvement and improve the company's green image. Some retailers are much more active than others. B&Q, Tesco and Sainsbury's are all partners in the 'Forum for the Future', which aims to be a clearing house for information on sustainable development. Membership is only available to those organizations who have demonstrated commitment to pursuing sustainable development (Green Futures, 1998, p. 4).

Environmental pressures on retailing

Traditionally most retailing texts' idea of the environment is confined to economic, social and political contexts. The idea that the physical environment might exert pressures is hardly considered (McGoldrick, 1990; Dawson, 1979; Davies, 1984). However, there has been a growing recognition that the physical environment and environmental issues do play a growing role in retailing (O'Brien and Harris, 1991; Kirk, 1991; Simms, 1992). In this section we look at pressures on retailers through a *green lens* using the familiar PEST framework (Figure 8.2).

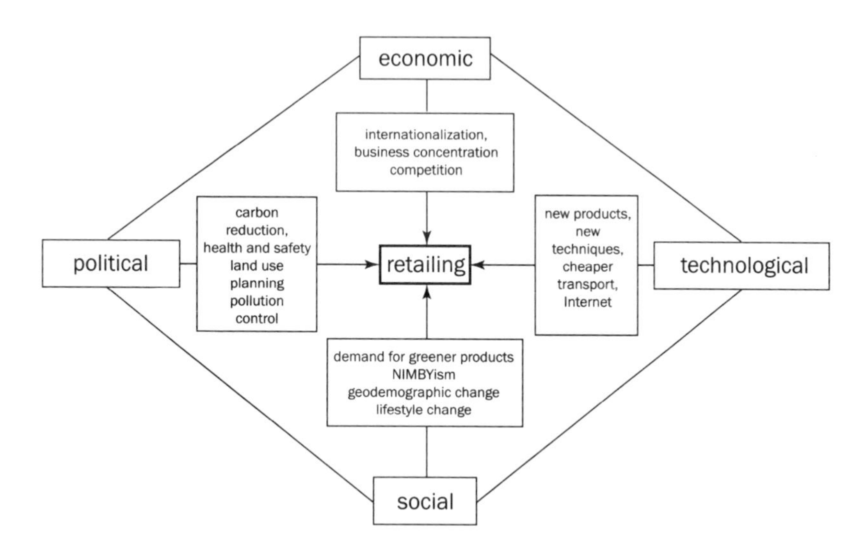

Figure 8.2 ***Environmental pressures on retailing***

Political pressures

Environmental issues have risen steadily on the political agenda in the past thirty years. Governments have committed themselves to goals such as carbon reduction, ozone layer stabilization and air pollution reduction. A major cause of pollution is road traffic and retailing contributes to this through delivery of goods and travel by customers. The growing tendency for out-of-town, or edge-of-town retailing has increased the amounts of car travel involved in shopping and the distances goods are transported has increased greatly. The result has been the identification of out-of-town retailing as a major causal factor in carbon emissions and hence has stimulated a desire by governments to reduce car travel. This has combined with other pollution concerns such as low-level smog with its implications for respiratory diseases. Pollution from retailing is a relatively new political idea for governments. Environment issues in the past were mostly to do with land use planning. This is still important, but is now combined with pollution as a motivation for controlling location of economic activities.

Retail developments have become large scale whether by individual stores or shopping centre companies and inevitably they have environmental impacts on the localities. It becomes a political issue when sufficient people are upset enough to protest, lobby representatives, and write to the newspapers. Councillors and MPs are sensitive to constituents' concerns, especially in marginal constituencies in the run-up to elections. Government attitudes in the UK have undergone several radical changes towards out-of-town retailing from prohibition up to 1979, to relaxation of restrictions in the 1980s and back to restraint in the 1990s. Political calculation plays a large part in government policies. If certain types of development become unpopular enough to influence the way people vote, restrictions can be expected.

Economic pressures

Retailing has become increasingly subjected to the pressures of internationalization (Treadgold, 1989). Deregulation and liberalization have made cross-border mergers and acquisitions, franchising and alliances increasingly common in retailing. When a retailer finds its home market saturated and is a unable to expand, some move abroad is likely. British retailers such as Marks & Spencer, Tesco and Sainsbury's have expanded into Europe, the Far East and North America during the 1980s and 90s. Sometimes they expanded by acquisition. For example Marks & Spencer bought Brooks Brothers in America. Sometimes a chain opens stores under its own name, for example, Marks & Spencer in France. What are the environmental implications of internationalization?

1 Lengthening the supply chain and increasing transport with its associated pollution.
2 Exposure to new and often variable environmental regulations. Companies may find that there are stricter pollution laws than those they are used to at home. There may also be different planning procedures and more stringent requirements for environmental reporting. Companies which operate internationally find it easiest to conform to the highest specifications demanded and then apply them uniformly, even if not legally required in every country.

3 Business concentration means that the biggest operators have steadily more influence in the trade and have proportionally more environmental impact because:
   (a) they are bigger targets for government pressure for action;
   (b) they have a higher public profile and need to demonstrate green credentials;
   (c) they are able to act as exemplars for small firms.
4 Competition: as product, prices and services converge, retailers seek any possible differentiation to attract customers. Trading on health, ethics and environment have been three ways to appeal to consumers, particularly to the ethical well-off customer. Promoting a store as green is good business. However, it is worse than useless to simply employ greenwash and pretend to have gone green without having actually done so. The resulting backlash to a firm could prove damaging in the extreme.

Social pressures

Retail stores provide a focus for community shopping areas where people often meet and socialize. Neighbourhood shops, particularly Post Offices, are a lifeline for the elderly (Granville, 1998). The decline of the small shop, especially the demise of the corner shop or village shop, has impoverished the social environments of the poor and less mobile sectors of society. The connection between the rise of the out-of-town shopping centre and the decline of the small local shop is obvious. Social changes in lifestyle with greater segmentation of consumers into numerous niches has meant that many shoppers are forced to travel long distances for shopping because there is no local choice. In inner-city areas the dearth of cheap food shopping has led to concerns about poorer people's health. Initiatives to improve diets and to subsidize community shops or box systems have begun (Smith *et al.*, 1998; Green Futures, 1999). Demand for green products is one obvious environmental pressure on retailers. Food, paper products, cleaning materials, cosmetics and pharmaceuticals, batteries and petroleum products have all been subject to green consumer demand. Durable goods like cookers, freezers and personal computers have begun to appear with green specifications. DIY products nearly all have environmental implications. Much of the timber in the major DIY chains like B&Q and Homebase are from sustainable forests with the Forestry Stewardship Council approval. This scheme in partnership with the WWF, aims to meet the concerns about the depletion of tropical hardwoods. Although timber is an obvious wood product, others such as hardboard, wallpaper and MDF are also made from trees. They are increasingly being scrutinized for their environmental implications. Paints, solvents and wood preservatives frequently contain volatile organic compounds (VOCs) and can be hazardous to the user as well as the general environment. Garden centres are often part of DIY stores and environment issues here relate to pesticides, herbicides, peat from unsustainable sources and bulbs from the wild.

Geodemographic change has been a major factor in store locations. The general decentralizing of the population from cities to suburbs to rural fringes has been part cause, part consequence of out-of-town retail shifts. The rural urban fringes are generally populated by the better off, who, although they are more than happy to patronize existing out-of-town stores, often display Nimbyism in the opposition to any new ones.

Technological pressures

These range from new products to new technologies for preservation and storage, in transport and increasingly, the Internet. The environmental effects of some of these are more obvious than others. Transport has become relatively cheaper and this has allowed global sourcing for supermarkets. While the distances involved in transporting beans from Guatemala, kiwi fruits from New Zealand and orange juice from Florida to Europe are substantial, little thought is given to the cumulative transport costs in environmental term. The Germans, for instance, consume 1.5 billion litres of orange juice annually which requires in turn 40 million litres of fuel oil and results in 100,000 tonnes of CO_2 emissions to get the juice to the table (von Weiszacker *et al.*, 1998, p. 121). Many products require inputs of a variety of materials and the transportation of those materials to assembly sites can total surprisingly large distances. In the case of strawberry yoghurt manufactured in Stuttgart, over twenty different materials ranging from strawberries to aluminium and plastic, travel a total of 3,500 kilometres. In addition to this journey, there is the supply of materials to the suppliers before they go on to the manufacturers. This adds a further 4,500 kilometres. So in all, the parts of the average yoghurt pot travel a total of 8,000 kilometres with all the implications for energy use and pollution that this entails (ibid. p. 117–19).

Environmental opportunities for retailing

Environmental opportunities for retailers can be broken down into six categories (see Figure 8.3).

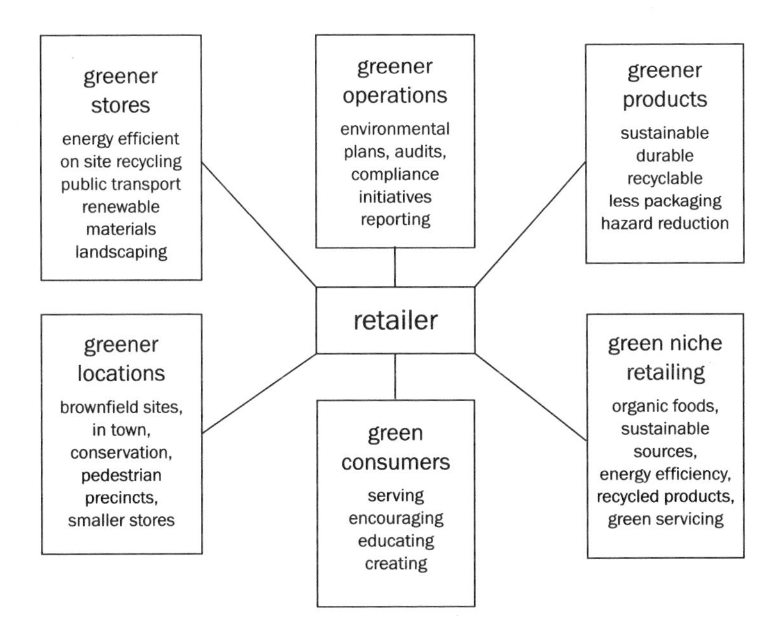

Figure 8.3 ***Environmental opportunities for retailing***

Greener operation

Whether motivated by cost-cutting, PR or a genuine effort to achieve sustainability, retailers can adopt a variety of green operations. Like all businesses, they have to comply with any environmental regulations, but those who are really committed to sustainability will go further than mere compliance. A complete environmental audit and management strategy of the kind used by Sainsbury's or B&Q brings the environment into every facet of the business. To be successful, environmental considerations need to be integrated into all aspects of the business. Reduction of waste, for example, can involve direct action on the part of the store and consumer action facilitated by the store. The retailer needs to act as an educator and encourager for the consumer. Programmes stand more chance of success if they are easy to perform, take little effort, do not cost consumers any more, and give a 'feel good' feeling as a reward.

Greener stores

Many retail stores built in the 1970s were often cheap-looking sheds of low architectural merit that were simply weatherproof enclosures for the retail activity inside them. By the 1990s the attitudes to stores had changed to make them more attractive buildings both inside and out. Shopping centres have become architectural showpieces. Bluewater, for example has re-creations of regional architecture inside its malls, ironically replicating the appearance of the traditional high street. The largest mall in the world at West Edmonton, Canada, incorporates ice rinks, water slides, aquaria and sculptures, as well as hotels, restaurants, cinemas and of course shops. Older shopping centres like Brent Cross, originally built in the early 1970s, have been given 1990s' facelifts to compete with the newer complexes. Energy efficiency, on-site recycling, renewable materials and landscaping are now the norm for new developments. Bluewater is actually built inside a disused chalk quarry.

Greener locations

One of the chief criticisms levelled against out-of-town retailing is that it leads to the simultaneous despoliation of both countryside and town. Greenfield sites in rural areas swallow up countryside, and generate traffic pollution in formerly tranquil areas. They also lead to the decay of urban shopping areas which become full of vacant sites, run down and prey to vandalism and crime. It was hardly surprising that restricting out-of-town stores became an easy target for politicians. At the 1994 Conservative Party conference, Environment Secretary, John Gummer won a standing ovation by declaring that there would be 'no more giant supermarkets desecrating the English countryside'. The incoming New Labour Government of 1997 essentially confirmed this change in policy and extended it to a preference for expanding development on brownfield sites.

As well as out-of-town developments being restricted, there are other environmental aspects of locations to consider. Shopping centre developments in town centres now pay much more attention to the environment (Box 8.2). Numerous urban shopping areas in European towns and cities have been pedestrianized, for example Copenhagen, and others have considered it, such as London's Oxford Street. Conservation of the urban heritage was once seen as preservation of landmarks of historical and cultural value. It has increasingly been seen as an asset to assist in urban regeneration by stimulating tourism and in differentiating shopping centres. In some cases historic buildings have been incorporated into store fabrics, for example, a redundant church in Wolverhampton now forms a Sainsbury's store main entrance (Larkham, 1996, p. 240). In West London the redundant Hoover building, a classic 1920s' Art Nouveau structure, has been incorporated into a Tesco superstore.

Greener products

Retailers do not, by definition, manufacture products, although some may have manufacturing siblings in the same group, for example, Boots the Chemist. However, large retailers have enormous buying power and can strongly influence or even dictate to their suppliers exactly what they want. They therefore have the power, should they choose to use it, to demand greener products. A greener product should be from a sustainable source, durable, capable of recycling, have minimal packaging and pose minimal hazards to the environment in production, transport, sale, use or indeed, after use. The case of CFC-propelled aerosols is a case where consumer concern led retailers to demand CFC-free aerosols from manufacturers. B&Q demands assurances from suppliers that timber is from certified stock and Iceland Frozen Foods took unilateral action to ban GM products from its range. This shows how action by a single company can act as a catalyst to others. Iceland's sales went up following its decision on GM foods and this led other food retailers to phase out GM products.

Green consumers

The idea of the green consumer dates from the 1980s with Elkington and Hailes' *Green Consumer Guide*. The idea that individuals can make a significant contribution through everyday purchasing patterns is a familiar one (Porritt, 1990; Seymour and Giradet, 1987). O'Brien and Harris (1991, p. 232) say that the aim of green consumer movements is: '[to] use the power of competitive free markets to redirect commercial activities from environmentally destructive forms of production and consumption into forms that sustained the planetary resource base while simultaneously maintaining the comfort of Western lifestyle'.

Box 8.2

The Chimes shopping centre development and the environment

The Chimes shopping centre, Uxbridge, West London, developer Capital Shopping Centres – the company which built and manages Lakeside shopping centre in Essex, one of Britain's largest. Development period 1998–2001; 420,000 sq feet enclosed shopping centre adjoining the High Street. Anchor stores 102,400 sq feet, Debenhams, 75 other stores including Boots and BHS, 9 screen cinema. Market valuation of £237 million

1. Construction: Excavation taking place on the side of a hill so that the three level centre only appears two levels high from the High Street to keep in character. 72,000 cu metres of material excavated and removed. 1,700 piles sunk by 80 ft piling rigs chosen as the quietest on the market to avoid noise pollution.
2. Refurbishment of historic buildings: Incorporation of restored and cleaned Victorian and early 20th century shops into the centre. Existing buildings used during construction phase as site offices and information centre to avoid building unsightly huts.
3. Chippendale Square: Specialist restaurant and café area on a new piazza focused on refurbished listed building Fasnidge House and a new timber-framed building in traditional style built using reclaimed timber from demolished buildings.
4. Vernacular style façade: New shop fronts on the High Street side will be built in the local traditional style to complement the existing style and architecture.
5. Light and energy: Glazed roof running the whole length of mall to admit natural light to help reduce lighting and energy and to create a more pleasant ambience.
6. Public transport access: As well as 1,600-space car park, public transport is well catered for. One entrance to the mall will be from Uxbridge Tube station and the bus station is nearby. New bus stops will be put in adjacent to other entrances.

Source: CSC information centre and field observation, March 1999.

To eco-radicals the phrase *green consumer* is an oxymoron, in their view, one cannot be green and a consumer. While this has a purist logic, it also has no chance whatever of becoming the lifestyle of choice of any but a tiny proportion of the population. It also has no appeal to retailers since their viability is not served by a cessation of consumption.

We have already considered some aspects of the green consumer phenomenon in Chapter 5 and we will not repeat them here, instead we shall consider what the retail business could do to serve this market. Retailers could simply respond to demand for green products but this would be slow and probably unprofitable. In other spheres, retailers are not passive responders to demand. Fashion retailers do respond to demand but only after it has been created by designers, advertising and media exposure. In the same way with green products, retailers need to create a demand by encouraging consumers to become greener by educating them to the benefits. Organic foods are largely promoted on the basis of the health benefits to the consumer with environmental friendliness to the countryside as an added bonus.

A problem of estimating the significance of the green consumer is that the idea is more frequently espoused than put into practice. Just as 65 per cent of the UK population will describe themselves as Christian in surveys, but only 11 per cent actually go to church once a week, so many people claim to be green but don't actually practise green consumerism (Nicholas, 1999).

Green niche retailing

Some retailers market themselves as being green – the Body Shop is a good example. It tries to source products from the Third World in non-exploitative ways in order to help producers and the environment. Cosmetic products are not tested on animals and make use of natural ingredients, packaging is minimal, a refill service is offered and there is no advertising. The Body Shop also promotes education in its shops (O'Brien and Harris, 1991, p. 163). It has had an effect on other retailers, for example, stimulating mainstream chemists like Boots to stock more naturally based products. More significant is where a large chain decides to incorporate green products and green planning into its mainstream operations. A good example is Sainsbury's environmental policy. When larger companies begin to become greener then there is more chance of its widespread adoption.

Summary

The all-pervasive nature of the tertiary sector in advanced economies, especially in cities, means that its environmental effects are widespread, even though individual service business may have a relatively small impact compared to counterparts in the primary and secondary sectors.

Retailing is an industry which has important environmental impacts and there is rising concern about the role of large superstore chains and out-of-town shopping centres in particular. Retailing experiences a range of environmental pressures and in a business that depends very much on consumer perception, increasing efforts are going into establishing green credentials. Retailers and shopping centre developers are beginning to incorporate environmental factors into their plans and operations. Many are also beginning to realize the potential for capitalizing on the opportunities presented by the green consumer, though there are dangers of superficial greenwash alienating both green consumers and the mainstream customers if there is no substance behind the rhetoric. Some retailers have taken opportunities offered by the environment and some large firms have begun serious environmental auditing and planning. The green consumer market is identifiable but fragmented and partial. It needs to be nurtured and developed if green retailing is to become more mainstream. It will have done so when consumers do not say: 'Why should I buy this just because it is green?', but 'I will not buy this because it is *not* green.'

Further reading

For a general review of the scope of the service sector D.G. Price and A.M. Blair, *The Changing Geography of the Service Sector* (London: Belhaven, 1989) provides an introduction. The main subject of this chapter, retailing, has a number of books worthy of consultation among which C. Guy, *The Retail Development Process: Location, Property and Planning* (London: Routledge, 1994), M. Kirk, *Retailing and the Environment* (Harlow: Longman, 1991) and R. Bromley and C. Thomas (eds), *Retail Change: Contemporary Issues* (London: UCL Press, 1993) are recommended. A useful source of retailing facts is Nielsen Ltd (annual series) *The Retail Pocketbook* (Henley-on-Thames: Nielsen/NTC Publications). Sainsbury's *Sainsbury's Environmental Report 1997/8* (London: J. Sainsbury, plc 1999a; available Online; http: //www.j-sainsbury.co.uk/environment) is a good example of a major retail firm's environment policy with a very comprehensive and easy to navigate website. For those who wish to examine other aspects of the tertiary sector, B.H. Massam, *The Right Place: Shared Responsibilities and the Location of Public Facilities* (Harlow: Longman, 1993) looks at the question of public services such as health care and waste transfer stations.

Discussion points

1 How can retail stores make their activities more environmentally friendly?

2 What are the environmental impacts of a major retail grocery chain?

Exercises

1 Box 8.2 shows how a developer has tried to incorporate environmental considerations into the building of a new shopping centre. For any shopping centre you are familiar with, evaluate how far it has addressed environmental needs.

2 Sainsbury's environmental impacts are shown in Tables 8.4, 8.5 and 8.6. Consult Sainsbury's website (www.j-sainsbury.co.uk) to read the whole Environmental Report. Compare this report with any other supermarket chains such as ASDA, Tesco or Safeway. Which company has done the most to incorporate environmental concern?

3 How green are the stores? For any stores of your choice make a quick visual estimate of the shelf space devoted to green products (e.g. how many shelves of organic food are there compared to the total food shelves?). What do they stock, where does it come from? Compare different companies' offerings. Retailers are often upset at people making notes in their stores (fearing spies from the competition), so if you want to do a full-blown survey, ask permission from the retailer, otherwise it is best just to make mental notes for rough estimates.

9 Environmental business

- **Conservation and business**
- **Business methods for a conservation organization**
- **The conservation multiplier**
- **Case study of the English country house: conservation or commerce?**
- **Environmental damage limitation and repair businesses**
- **Environmental business services**

> The RSPB estimates that the annual spending of British birdwatchers is now about £200 million.
>
> (P. Brown, *Guardian*, 20 August 1999)

> Heritage is now big business and it is sometimes criticised as being too commercial, however, as the past 40 years have proved, particularly at Beaulieu, well managed tourism can do more to conserve the heritage then any other source of funds.
>
> (Lord Montagu of Beaulieu, 1992)

Environmental businesses can be seen as falling into one of the following three categories, each with its own characteristics:

1 Conservation businesses which seek to preserve or protect an aspect of the environment, for example, wildlife, landscape or heritage. Dwyer and Hodge (1996) used the term CART – conservation amenity and recreation trusts. Our definition of conservation business is a little wider and includes both public, voluntary and commercial organizations.
2 Environmental damage limitation or repair businesses: EDLRB – this will include businesses concerned with waste management, pollution control, recycling or remanufacturing.
3 Environmental business services. These businesses are concerned with servicing other businesses in the environmental sector, other non-environmental businesses, or the general public, for environmental

purposes. This includes the environmental media, environmental consulting, ethical investment trusts, environmental business clubs and associations, and public/private environmental partnerships.

It might be supposed that a sustainable business could fall into the environmental business category. Any business in any sector could have sustainability as an objective and need not necessarily be one of the above specific environmental business types. Sustainability is a way of conducting business rather than being a particular type. It seems at present to be an ideal, which although widely discussed is rarely practised. Most commentators see it as necessary for the very survival of civilization, but despite this general endorsement it has not got far in practice (Brown and Flavin, 1999; Welford, 1997). In this chapter we shall be concentrating on examining the three types of environmental business, as they have not been considered in this framework before, while sustainable business is a very widely discussed concept amply covered by the literature (Pretty, 1998; von Weizsacker *et al.*, 1998; Welford, 1994).

Conservation and business

Some businesses have turned aspects of conservation into a commercial activity. We have identified three types:

- wildlife-based, e.g. RSPB;
- landscape-based, e.g. Becky Falls in East Devon;
- heritage-based, e.g. Beaulieu Estate.

A distinguishing feature of conservation businesses is that they have become more commercial in recent years and we must investigate why this has happened. Up to the late 1970s most conservation was non-commercial in ethos and practice. Government agencies or charities managed their charges without attempting to make money from them in any significant way. Admission charges to beauty spots, historic houses and ancient monuments, or wildlife sites were essentially to cover costs and were generally low and for many places were free. Countryside recreation in particular was regarded as a free good. The reasons for this were:

1 Low demand since leisure time and mobility were considerably more limited than today. Leisure activities were more focused on traditional urban venues such as cinemas and spectator sports.

2 Minimal pressure on the environment and therefore little damage to the fabric of sites and consequently little need for repair.
3 Low expectations. People did not expect much of a wildlife or heritage site except to see it.
4 Basic presentation technology – glass case, hand-written label at best, with no interactive multimedia.
5 Consensus politics – agreement that the state should have a role in planning and conservation, 1945 to 1979. Governments of either party did not question the fundamental role of the state in looking after the environment heritage and providing sufficient funds for its maintenance.
6 Limited view of what should be conserved by the state or trusts (preservationist ethos).
7 Perception that things were basically unthreatened and that private owners were stewards.
8 Limited number and sizes of non-profit-making organizations (NPO). Even what have become large organizations such as the National Trust and the RSPB were relatively modest-sized organizations at this time. Many other NPOs did not come into existence until the late 1970s.

The Thatcher/Major Conservative Governments of 1979–97 were marked by an ideological approach unusual for the UK. The idea was simple: reduce the size and role of government and let the markets decide the value of everything. The private sector should be allowed to expand as far as possible and the public sector should be reduced to the bare minimum. The results were:

- a reduction in state expenditure therefore less money to go round;
- privatization of a range of government functions resulting in the quangoization of many government agencies;
- very widespread deregulation to free the private sector from the burden of red tape;
- extending the business ethic and the methods to public sector agencies. As Will Hutton (1996, p. 217) said, ‘The extension of the market has gone well beyond the formal institutions of the welfare state. It now extends into the very quick of society.’

So pervasive had the ideology become that the incoming Labour Government of 1997 not only continued, but also extended the policies of privatizing public bodies. According to the Prime Minister Tony Blair the public sector was still not entrepreneurial enough (*Guardian*, 7 July 1999, p. 2).

At the same time the environmental sphere saw an upsurge of public interest. A growth of membership of environmental organizations corresponded to a vast increase in visits to the countryside, coast and historic buildings. This was partly due to the fruition of earlier government policies in the 1970s, which sought to encourage recreation and sport (Sport for All). Local authorities had been encouraged to develop outdoor recreational facilities and in particular the concept of the Country Park was designed to entice people into the countryside, albeit in a controlled way. There was an increased perception of damage being done to the environment by commercial interests, especially farming, causing damage to landscape and wildlife. Stricter environmental regulations emanating from the European Union which were supposed to be adopted by Member States, but given Tory antipathy to Europe, led to many highly publicized arguments and even more publicity for the issue. Finally, several critical books on the environment played a major part in stimulating the debate, for example, those by Shoard (1982), Porritt (1990), Rose (1990), and Harvey (1998). The outcome in the 1980s was therefore a drastic change in the attitude and practices towards conservation. On the one hand, a growing demand for more to be done, but, on the other, a retreat by government from the welfare state prompted by a desire for lower taxation as a vote winner.

There was a growing demand from the public to visit wildlife, landscape and heritage sites. Information and communication technology has been revolutionized and the standards of presentation have improved. Visitors expect more spectacle. There was in consequence more pressure on site managers because of more visitors and this led to more environmental pressures where erosion of sites became a serious problem. However, static public expenditure, which in effect meant reductions when increased demands and inflation were taken into account, meant a pressing need for more funds from some source or other. There was also increased statutory demand to conserve a wider range of landscape, wildlife and heritage, for example, the Wildlife and Countryside Act 1981. There was an infusion of the market ethos into government agencies, such as the Next Steps Agencies e.g. the Royal Parks Agency, Historic Royal Palaces Agency and Welsh Historic Monuments (CADW). Pressure on non-profit organizations to generate more funds for conservation increased as membership, the number of sites and public expectations all increased.

The results of these changes were:

1 more enhancement of the conservation offering, more value-added sites with interpretation facilities, and multimedia tours, re-enactments and special events;
2 extension of merchandising, for example, the National Trust extended its shops from being only on its own properties, to around forty town sites. The National Trust had become a brand in its own right;
3 more business orientation of NPOs with active searches for moneymaking opportunities;
4 more business orientation of government agencies;
5 more of a role for the voluntary sector in conservation due to the expansion of NPOs, e.g., the Woodland Trust acquired its thousandth wood by 1999;
6 a greater willingness of the public to pay a fee as evidenced by growing subscriptions donations, legacies, and admission charges for things formerly free.

We have coined an acronym CANPO – commercial arm of a non-profit organization, to describe the business activities of NPOs (see Figure 9.1). They may be described by the NPOs themselves under a number of headings such as membership departments, business departments, development departments, planning and research, marketing, etc. It is possible for any NPO to have a commercial activity, for example, Oxfam and Barnardo's Children's Homes are examples of non-environmental charitable organizations that raise money through business activities, such as Oxfam shops, and Christmas catalogues. Such activities are therefore not exclusive to environmental NPOs but environmentalist CANPOs often are distinguished by making money from activities pertaining to specific sites, for example a wildlife reserve or an historic property. There are essentially three ways to pay for conservation objectives (Figure 9.1):

1 non-commercial: through state agencies via taxation which predominated before 1979 and still exists but in reduced form; or through charitable trusts which derive their funds from the donations, grants, legacies and membership fees.
2 commercial: whose aim is to make money, i.e., conservation is mainly cosmetic. Commercial sites can make valuable contributions to conservation but full commercial exploitation, while paying lip service to conservation can result in damage to the environment.

3 CANPO I and II: CANPO I where raising finance for conservation objectives is the primary aim and there is no compromise in environmental integrity. In other words, the preservation of the site will always be a higher priority than raising money from business activity. CANPO II where the business activities become increasingly important to the extent that the tail wags the dog. There is a danger for NPOs where there is a temptation to maximize revenue by an over-extension of business activities, unless the CANPO is careful. It is vitally important for the aims and image of the NPO that business activities are subordinate to their conservation objectives. An excess of commercial zeal could rebound by deterring conservation-minded members of the public from supporting the organization.

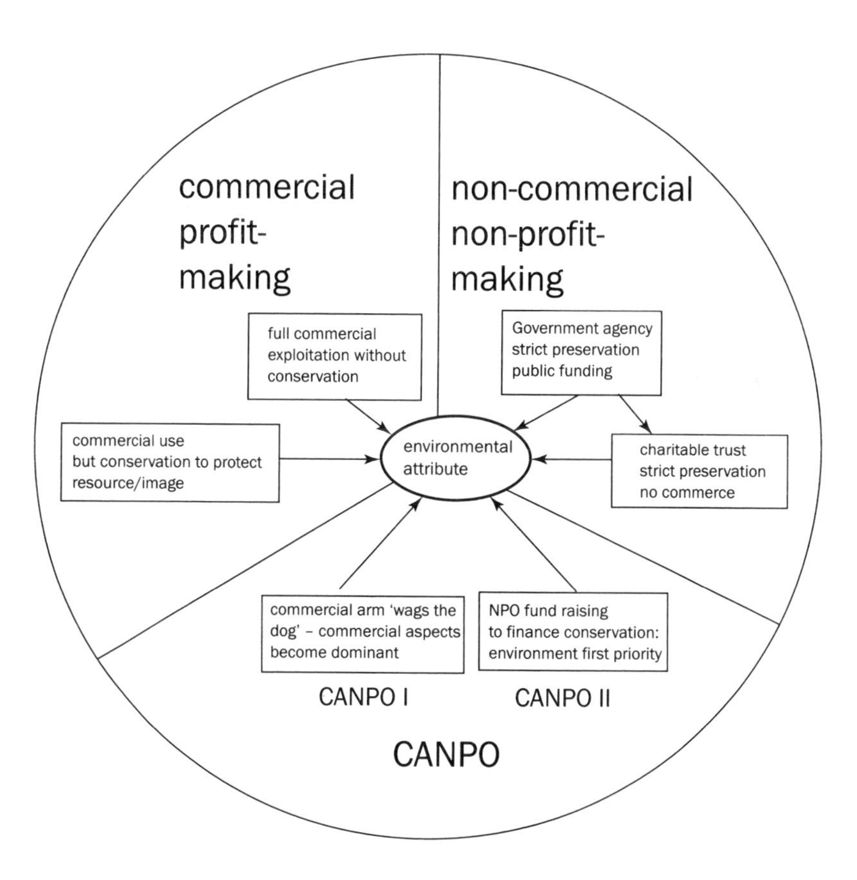

Figure 9.1 ***CANPOs – The Third Way of conservation?***

Business methods for a conservation organization

There is a wide range of business methods potentially available for CANPOs (Box 9.1.) Dwyer and Hodge (1996) estimate that there are 126 UK CARTs (not quite the same as CANPOs but similar) having an annual turnover in 1990 of over £110 million. This was 50 per cent higher than the total public money given to all official conservation bodies. The CARTs covered 2.7 per cent of the UK land area. The largest four bodies, the National Trust, the National Trust for Scotland, RSPB and the Wildlife Trusts accounted for 80 per cent of this area (ibid., p. 55). NPO conservation organizations are therefore significant owners of land and significant businesses as well as being guardians of the environment. Both voluntary and official organizations have as their objectives safeguarding the aspect of the environment in their charge, for example, birds for the RSPB, woodlands for the Woodland Trust, historic built environment for English Heritage. Their primary function is conservation, but conservation is not cost free and since the Thatcher revolution changed attitudes to public funding apparently permanently, some of the funding for conservation has to be raised by business activities.

Box 9.1

Business methods for conservation organizations

- advertising
- charging for admission and or car parking
- value-added activities, e.g. rides, films
- onsite or theme-specific merchandising (souvenirs, video, books)
- special events (concerts, historical re-enactments)
- affinity cards (percentage of transactions)
- catering
- hosting (conferences, dinners)
- sponsorships/partnerships
- film location charges
- membership subscriptions/legacies/donations/raffles
- tours/holidays
- mail order catalogues
- renting property
- investments

The ways of raising money are highly diverse and we do not intend giving a detailed discussion of all of them which would be space-consuming and tedious. We shall instead highlight some of the more interesting methods and give some examples of CANPO activity. A review of the selection of conservation NPO publications in 1998–99 showed the extent to which English CANPOs use a wide variety of methods to raise funds (Table 9.1). The first point to note is how each of them employ a very wide range of methods. The chief difference lies between those bodies which own property to which the public is admitted (NT, WT, RSPB, EH, RBGK) and those who do not (WWF). A further difference is that the National Trust, which has by far the biggest estate, is also a major landlord and earns significant rental income. It is possible that some organizations do more things than are recorded simply because of the time frame of the survey. Regardless of the completeness of the record, the most striking feature is how similar all these bodies are in their fund-raising activities.

Table 9.1 ***English CANPOs' activities***

	Organization					
Activity	*National Trust*	*Woodland Trust*	*WWF*	*RSPB*	*English Heritage*	*RBGK*
rental income	yes	no	no	no	no	no
membership subscription	yes	yes	yes	yes	yes	yes
admission charges	yes	no	n.a	yes	yes	yes
merchandising	yes	yes	yes	yes	yes	yes
fund raising events	yes	yes	yes	yes	yes	yes
sponsors/partners	?	yes	yes	yes	yes	yes
tours/holidays	yes	no	no	yes	yes	yes
affinity cards	yes	yes	yes	yes	yes	yes
mail order catalogue	yes	yes	yes	yes	yes	no
film location income	yes	no	no	yes*	yes	no
Lottery grants	yes	yes	?	yes	yes	yes
volunteers	yes	yes	n.a	yes	?	no

Sources: *Birds*, *Kew*, *Heritage Today*, *WWF News*, *Broadleaf*, *National Trust Magazine*: all editions for each periodical from August 1998–August 1999.

Notes:
WWF = World Wide Fund for Nature
RSPB = Royal Society for the Protection of Birds
RBGK = Royal Botanic Gardens, Kew
* RSPB makes films of its own for commercial showing
? indicates no evidence in publications examined
n.a. 'not applicable'

The Woodland Trust exemplifies many of these points. It was founded in 1972 by Devon businessman Kenneth Watkins and has had a strong business input from the start. The Woodland Trust exists to protect native British woodlands and to extend the area of woods by new planting of native species. It has a strategy of acquiring woods and land through ownership. This means it needs finance to buy property as well as to plant and maintain woodlands. It has experienced rapid growth of both membership and woods. It keeps in touch with its members through a regular newsletter *Broadleaf*, a directory of woods and frequent appeals for donations, merchandise catalogue, raffles and special events. Unlike many CARTs with property, it does not charge for access to its woods which are also freely open to non-members. Table 9.2 shows a breakdown of the sources of the Woodland Trust's income amounting to £14 million in 1998. It can be seen that the biggest shares come direct from the public in the form of membership subscriptions, donations and legacies (41 per cent) and indirectly from the public in the form of Lottery grants, and other grants (32 per cent).

Merchandising is relatively small at 2 per cent and this is much less than the contributions of merchandise to the National Trust (Clayton, 1986; Dwyer and Hodge, 1996). However, the National Trust has much more opportunity with not only a shop in almost every property, but also a network of National Trust shops in historic towns. Similar proportions come from investments and woodland management, for example, sales of timber. One of the most interesting features of the Woodland Trust fund-raising is the role of sponsorship and partnerships. Here, businesses as diverse as Microsoft, News International, Castle Cement, Jaguar Cars, Canon, Honda UK, Unilever, British Nuclear Fuels, Zeneca and the Royal Bank of Scotland have all contributed in some way. The Woodland Trust lists eight different forms of sponsorship/partnership in its 1997 Review (Woodland Trust 1997b, p. 15):

Table 9.2 ***Woodland Trust sources of income***

Income source	%
membership subscriptions	10
public donations	15
landfill tax credits	7
donations from trusts	3
donations from companies	4
legacies	16
Lottery grants	15
other grants	17
investment income	3
woodland management income	5
donated land	3
merchandising and lotteries	2

Woodland Trust total income in 1998 = £14.65 million

Source: Woodland Trust, 1998 *Summary of Accounts*, Grantham.

1 Funded woodland purchase: seven companies, twenty-eight others.
2 Contributed to woodland purchases and initial management: five companies, thirteen others.
3 Woodland management conservation: ten companies, eleven others.
4 Donations: seventeen companies, nine others.
5 Support through landfill tax credits scheme: fifteen companies, three others.
6 Corporate members: thirty companies.
7 Gifts in kind (e.g., raffle prizes, materials and equipment): eleven companies.
8 Partnerships: eight companies.

Altogether, eighty-seven separate companies are listed (some appear in several categories). Most of the categories of partnership or sponsorship are self-evident, but two items, the landfill tax credit and partnerships merit further elaboration. The landfill tax was introduced in 1996 and has been described as the UK's first real attempt to green the tax system (Howes *et al.*, 1997, p. 1). It is designed to encourage a reduction in waste disposal by landfill and imposes a levy on every tonne dumped. It is possible for landfill operators to claim a rebate of 90 per cent of their tax payments if they donate it to environmental trusts. The Woodland Trust describes these tax credits as an 'increasingly significant source of funding' (Woodland Trust, 1997b, p. 15). By 1998 landfill tax credits accounted for 7 per cent of the Trust's income. Partnerships are where there is mutual benefit for the Trust and its business partners through operating a variety of schemes. There were eight partnerships in 1997 where a business undertook an operation designed to raise money for the Trust. One example was Furnitubes International, which undertook a promotion linking sales of timber products to tree planting. Another was Continu-forms Holdings, a printing company, which included tokens on its products, which if returned to the company, would be credited to one tree planting per hundred tokens in one of nine named woods. The Royal Botanic Gardens Kew raises money from corporate membership at £2,750 plus VAT per year, or £11,000 plus VAT for five years, or from Corporate Friends at £1,100 plus VAT. Corporate members can attend special tours and the evenings at Kew Gardens or Wakehurst Place. Companies enrolled include Smith Kline Beecham, Thames Water, Pearson plc. (Kew, 1999, p. 45).

A mutually beneficial business partnership activity is the affinity card. This is a normal credit card run by banks or financial institutions which is associated with a charity and usually has a design and logo of the charity prominently displayed, for example, the RSPB affinity card has a photo of a Kingfisher, chosen by the marketers because it came top of a popularity poll, while English Heritage card has a choice of ancient monuments such as Stonehenge. As Davies (1999) explains, the RSPB affinity card is a highly effective fund raiser, gaining the society over £3 million since its launch in 1989. The RSPB card is used by 100,000 people who automatically pay one-quarter of a percentage commission from transactions to the Society. Affinity cards are growing in popularity with an estimated 1,200 in the UK alone. RSPB is in partnership with the Co-operative Bank, which owing to its origins in the Co-operative movement, has a long tradition of ethical policies and attracts many charities as clients. The affinity card seems to display the characteristics of a win-win-win situation. Card users win because they are supporting their chosen charity, but they hardly notice a regular contribution as it is a very small percentage of the transaction costs and no special effort is needed to make donations. The RSPB wins because they get a regular long-term income at little cost, unlike other revenue-raising schemes such as appeals that require resources and organization for uncertain return. The Bank wins because although they donate a sum to the society, they have increased their customer base at low cost and those customers tend to be more loyal than the average customer and stick with their affinity card. Although virtually all charities now have their own affinity cards, the only conceivable problem is the limit on numbers of people deemed creditworthy who are prepared to take one out.

The conservation multiplier

Conservation businesses not only provide employment and income generation directly, but they also have a multiplier effect. In other words, their existence in an area gives rise to further business opportunities which would not exist if it were not for their presence. For example, National Parks exist to preserve and enhance the natural and cultural environment of particular designated areas. At the same time they seek to encourage visitors to the area for recreational and educational purposes. A strong theme in all National Parks is to balance the needs of conservation with visitor enjoyment. Unlike a nature reserve, where preservation for scientific purposes is the over-riding concern, National

Parks depend on public support for continued funding and increasingly for direct income. In their local area National Parks can act as a catalyst for business activities. In England and Wales the National Parks are in a weak financial position. The total state support amounted to £19 million in 1999, which is £7 million less than the Royal Parks in London received. Public funding only meets two-thirds of the National Parks' expenditures so they have to generate additional income from activities such as guided walks, admissions to interpretation centres, grants and sponsorships (Hall, 1999). All the National Parks are in remoter areas, usually with limited employment opportunities relating mostly to agriculture, forestry or mining, all of which have seen contraction in recent years. National Parks are popular tourist and recreational destinations and can therefore act as a stimulus to business by providing many different opportunities. A village store, for example, may only be viable because of its high sales to summer visitors to the National Park and this balances out low sales to local people out of season. The multiplier can extend well beyond the limits of the National Park; demands for guide books, maps, books, calendars, binoculars, boots, tents, rucksacks, waterproof clothing, etc., give opportunities to retailers and manufacturers. It is impossible to say what proportion of these things are specifically purchased because of a National Park visit, as many can be used elsewhere, but some undoubtedly are bought because of such trips.

A second example of the conservation multiplier relates to urban environments. Increasingly there is a premium being placed on locations with architectural and/or historic interest. Larkham (1996) notes that offices in conservation areas or listed buildings generally attracted a 10 per cent price premium over similar unlisted property outside a conservation area. Urban conservation has moved on from simply preserving rare or spectacular buildings to caring for the more mundane but nevertheless worthy structures in the everyday urban scene (Clarke, 1999). There is now a realization that urban regeneration can be aided by conservation. A good example of this use of conservation is the regeneration of Albert Docks, Liverpool. Old and new were blended to create an attractive environment that proved very successful. Some 800 new jobs were created and the Albert Docks are said to be the second most popular tourist attraction in the UK outside London with 5.8 million visitors per year (Pearce, 1994). The environment itself, whether natural or the product of human activity, can therefore be a powerful aid to increasing business opportunities in general.

Case study of the English country house – conservation or commerce?

English country houses that survive today were built, between the late Middle Ages and the Edwardian era, by rich owners for their own residence. Thousands survive today and are increasingly regarded as important elements of culture, history, environment and heritage. However, changing economic, social and political circumstances mean that few private owners can afford to maintain such structures purely for their own use. The decline of the aristocracy along with the increasing burdens of death duties, following the death of many heirs during the First World War, led to the country house being seen as a cost rather than an asset. Many were converted to private schools, mental hospitals, local government offices and during the Second World War temporarily to military and intelligence uses. In the 1950s began the development of the 'stately homes' industry that was designed to make money for the hard-up aristocracy. In the process the idea of 'heritage', that the private residences of the aristocracy with their works of art and landscaped gardens somehow belonged as a cultural inheritance to the whole nation, flourished and spread to an extraordinarily wide degree. By the end of the twentieth century heritage was big business and had spread to other constructions such as coal mines, derelict dockyards and prehistoric monuments.

Although the modern cult of the country house can be dated to Lord Montagu's opening of Beaulieu in 1952 (Box 9.2), stately homes had been open for public viewing in the 1840s–70s. Mandler (1997) shows that the early to mid-Victorian owners of 'olden time' houses (i.e. pre-1660 vintage) capitalized on the current fad for 'Gothick' (*sic*) romances by magnanimously opening their properties to the public. Their motivation was to bolster their flagging public influence and esteem which had been undermined by the 1832 Reform Act and the declining economic asset of agricultural estates. At this time charges were purely nominal, usually to cover the costs of guides. But from about 1886 there was an abrupt change. Owners became reluctant to open their houses to the increasingly anti-aristocratic public. Public attitudes were turning away from reverence for the past and towards modernity and progress. Leisure time pursuits at the new seaside resorts were far more popular and in the twentieth century at the cinema or at spectator sports like football or cricket. The modern interest in what is after all the accumulated possessions of the very rich and their ancestors, has grown

Box 9.2

The Gilt and the Gingerbread – Beaulieu: conservation and commerce

Lord Montagu of Beaulieu in his book entitled *The Gilt and the Gingerbread: Or How to Live in a Stately Home and Make Money* (Montagu of Beaulieu, 1967) describes his 'stately home' business. Beaulieu was one of the first to put conservation on a commercial footing. According to Lord Montagu (1992, p. 3): 'My main motive for opening my home to the general public in 1952 was to generate sufficient income to restore and conserve the historic buildings and landscape of the Beaulieu estate.'

The Beaulieu estate is best known for the National Motor Museum, but it also has Palace House (the 'stately home'), extensive ruins of a medieval Cistercian abbey, fourteen farms, a vineyard, 2,500 acres of woodland, pheasant shooting, Bucklers Hard Maritime Museum, a 110 berth yacht harbour on the Beaulieu River and a National Nature Reserve. The development of commercial conservation can be shown by the increase in visitor numbers and admission charges. In 1952, 70,000 people each paid 2/6d (12.5 pence), by 1992, 500,000 visitors were paying £6.80 each. Staff employed increased from 20 to 400 over the same period and the average visitor time spent went up from 1 hour 20 minutes to 4 hours 40 minutes. From casual beginnings the Beaulieu estate has developed into a carefully planned commercial conservation business. Beaulieu is widely regarded as a model enterprise and it received so many requests for advice from other historic houses and museums that it set up Ventures Consultancy to provide advice commercially.

Source: Montagu of Beaulieu (undated, post-dates 1992) *40 years at Beaulieu*, Beaulieu.

and in the process the estates and houses have become regarded as valuable national assets. Many have passed into public ownership in one way or another (Figure 9.2).

The State in its various guises controls some large and well-known houses. At one extreme, commercial considerations play no part at all. For example, Chequers, the country retreat of British Prime Ministers, is not open to the public for security reasons. It is maintained in immaculate condition at some cost for prestige reasons since it hosts visits from foreign dignitaries. Interestingly, some royal houses are open to the public when the Royal Family is not in residence and an estate like

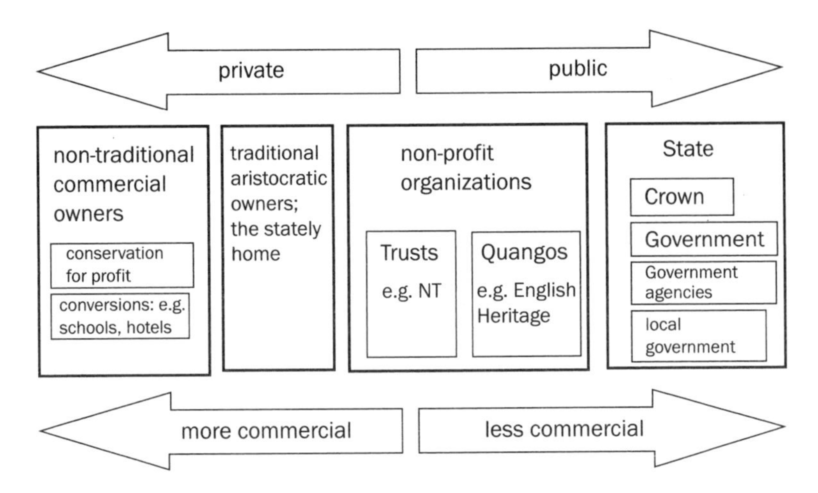

Figure 9.2 ***Profit and preservation: conserving the English country house***

Sandringham has similar attractions to those found in other stately homes, such as cafeteria, souvenir shops, garden centres, tours round the house, nature trails, etc. Several members of the Royal Family have publicly professed interests in conservation and since they are increasingly subject to the same commercial pressures as private owners, their estates are managed along conservation lines similar to National Trust properties. Indirect state involvement in country houses is through quangos such as English Heritage and the Historic Royal Palaces Agency. These levy admission charges and in the case of English Heritage have a wide range of associated fund-raising activities.

The National Trust is the best known of the NPOs managing country houses and puts the conservation of properties and grounds as its primary priority. It employs a wide range of fund-raising activities and its shops have spread beyond the National Trust properties themselves to town sites, so well known has the National Trust brand become. The National Trust rents out holiday apartments at its properties; normally these are the old estate cottages or converted stables, not the actual country house itself. A good example of this can be seen at Igtham Mote in Kent, where the former domestic buildings, which are themselves picturesque timber-frame structures overlooking the moated medieval manor house which is only open for visiting, are let for holiday purposes. National Trust properties vary in size but they do include some of the grand country houses like Waddesdon Manor, Montacute and Hughenden. Many of these big houses have been given to the Trust on the condition that the donor family is allowed to continue living in part of the property.

Quangos such as English Heritage normally manage properties where the owners have long since departed and the house is in effect a museum. Audley End in Essex and Osborne House on the Isle of Wight, are examples of very large and well preserved houses. Osborne illustrates how public agencies use commercial methods to raise funds for conservation. It was Queen Victoria's island retreat and after her death was given to the nation. English Heritage has undertaken an extensive conservation programme of the house and grounds. The house is a popular tourist destination and money is raised from admissions, catering, souvenir shops selling a wide range of Victorian themed merchandise from books and postcards, to jewellery and porcelain. We have already noted (Chapter 5) how English Heritage capitalized on the film *Mrs Brown*, which featured Osborne, not only from direct film location fees, but from the resulting free advertising which dramatically increased visitor numbers.

The Landmark Trust is a much smaller organization which deals with generally less famous properties which slipped through the net and would probably deteriorate or be otherwise demolished, for example, old windmills and country railway stations. Their approach to conservation is different in that the properties are renovated and then rented out for periods from two weeks to a year, but not for permanent residence. The idea is that the property should be used and the rents finance the conservation work (Landmark Trust, 1999, online).

The private country house sector can be divided into three main sections with the fate of the country house very different for each type:

Traditional aristocratic owners

The old aristocracy and gentry who have owned their houses for generations and still live there have entered the stately home business with varying degrees of enthusiasm and skill. The four modern pioneers are Lord Montagu of Beaulieu (Box 9.1), the Duke of Bedford at Woburn, the Marquess of Bath at Longleat and the Duke of Devonshire at Chatsworth. Woburn Abbey in Bedfordshire has over the years seen a diverse set of commercial enterprises in addition to the standard tour of house and gardens. These have included an antiques village of forty shops in the old stable block, a Safari park, hot-air ballooning, pop concerts, exclusive banquets with the family, and a nudist camp. The traditional owners' aims have been primarily to use commercial enterprise to raise funds to maintain their estates, which conventional enterprises such as farming and forestry could no longer achieve. Since they all continue to live in the houses, maintenance of their lifestyle is also an aim.

Conservation for-profit enterprises

In some cases the old owners of the country house have sold out to commercial concerns who then develop the attractions to increase tourism and revenues. A good example of this is Warwick Castle which was sold by the 8th Earl of Warwick to Tussaud's in 1970. Tussaud's is most famous for one of the biggest paying tourist attractions in London – Madam Tussaud's Waxworks/London Planetarium. It also operates a number of other major tourist attractions including the Alton Towers theme park, built around the core of the country house. Warwick Castle is a large imposing medieval structure that has been extensively restored. The new owners have capitalized on their asset by high quality state-of-the-art displays, for example, the Kingmaker exhibition that recreates one day in 1471 in the life of Warwick the Kingmaker. There are also demonstrations of falconry, jousting, minstrels, re-enactments, costumed guides and medieval banquets as well as the usual tours of apartments, dungeons, armoury and walls. The aim is commercial but without the well-conserved castle and grounds, the attraction would not exist. This is a good example of where conservation and commerce go hand in hand. Unlike, say, Euro-Disney, which is wholly artificial, Warwick Castle has at its core a genuine historic property.

The converters

Here the structure may remain but it has been converted to something commercial and unrelated to its past. Conservation is practised only so far as it is necessary to comply with regulations, for example, listed building status, or because it provides a commercial advantage. Thus, a country house hotel can charge a premium for the ambience and this may attract conferences and sporting events such as golf tournaments. Some country houses have been turned into expensive health farms, for example, Champneys in Hertfordshire or clinics for celebrities such as The Priory. Others have been turned into public schools, such as Ramsey Abbey near Peterborough, or residential apartments such as Shardloes near Amersham. Mentmore Towers was sold in 1977 to the Maharishi Foundation to become a centre for the Maharishi University and the Natural Law Party. All of these in effect ceased to be country houses in all but external appearance. If country houses do not attract some commercial use, they are unlikely to survive and will be demolished.

Environmental damage limitation and repair businesses (EDLRB)

EDLRBs are perhaps the most obvious and visible environmental industries. Any company whose primary purpose is to limit or repair damage to the environment could be classed into this group. This includes waste management, recycling, remanufacturing, pollution control and prevention. Some authors would classify renewable energy and water as 'green' industries but we have included them under the primary industries since both have a strong resource component. Few classifications are without anomalies and we are conscious that this particular grouping is far from perfect. Increasingly, as the requirements of business and environment converge, major companies in a variety of fields are setting up or taking over specialist environment damage limitation repair businesses. For example, Ford Motor Company bought a Florida automotive recycling company and plans to buy dozens of similar businesses with the aim of becoming the world's largest car parts recycler with a forecast annual revenue of $1billion within five years. All types of vehicles will be dismantled and the parts sold world-wide. The recycled content of Ford's own cars are planned to increase from 75 per cent to 90 per cent within the same period (SustainAbility, 1999, online).

Environmental problems may arise from an industry without any environment damage limitation repair businesses. Adverse effects on the environment are numerous and despite short-term profits, the industry may suffer from long-term unsustainability. Environment damage limitation repair business activity helps to remedy these problems and brings about benefits to the environment and for the client as well as economic benefits for the repair business itself. The kind of activities involved include clean technology, waste minimization, recycling, remanufacturing, energy efficiency and sustainable resource use. Short-term costs should be balanced and may frequently be outweighed by long-term savings.

Waste is a growing problem in developed consumer societies. In the UK 402 million tonnes of waste is generated annually (Brown, 1992). Disposing of waste is an increasingly difficult and costly problem – simply finding enough landfill sites within economic range of cities is becoming more difficult. International conventions on marine pollution further restrict disposal. Traditionally waste disposal was a publicly funded function with local government being the primary collection and disposal agency. In the 1970s in the USA it became apparent that waste collection and disposal were becoming a growing burden on the taxpayer. According to Gandy (1994, p. 12), the forty-eight largest cities in the USA were spending half of their environment budget on waste and the cost of collection amounted to over 70 per cent of that. Similar trends were evident in the UK and the political changes of the New Right in both the USA and the UK in the 1980s led to radical changes to the way waste was managed. The private sector has moved into waste management. In the UK, local government was obliged to institute competitive tendering for many of its services. Privatization of water and energy utilities sometimes led to a branching out by these companies into waste management. Smaller firms have been squeezed out by bigger concerns. Waste management and related services in the UK have an annual £3.5 billion turnover (ESA, 1998). Waste management includes collection, treatment, disposal, reuse, recycling and recovery of waste. It also includes specialist equipment and environmental consultancy services. An example of the complexity of environmental service businesses is shown by Général des Eaux, the French utility company which describes itself as one of the world's leading environment service providers. In the UK there are over 40 Général des Eaux subsidiaries. Best known for their water supply companies, they also have specialists in eighteen environment repair damage limitation businesses, including Onyx Waste Management Services and the ETS group, which provide energy and facilities management (Business in the Environment, 1999, online).

Recycling is one of the best known and popular green activities. The European Union has directed that 15 per cent of plastic packaging should be recycled by 2000. As we saw in Chapter 5, many members of the public practise recycling. Local authority levels of recycling in many cities are frequently low, mostly under 10 per cent and eight London boroughs achieved less than 4 per cent recycling of collected waste. These boroughs blamed the weakness of the secondary materials market, the need for more 'on street' collection facilities, restriction on local government expenditure and the high cost of kerbside collection as reasons for the low total (Gandy, 1994, p. 58). The business approach to recycling seems more successful than the purely municipal one. Reprise Limited, acquired in 1997 by PVC Group, operate a partnership with local authorities and voluntary groups to recycle plastic bottles. They have set up the first European plastic bottle reprocessing plant to demonstrate the feasibility of plastic bottle recycling. They have links with local authorities, soft drinks manufacturers and RECOUP (recycling of used plastic containers), a non-profit-making organization dedicated to recycling plastics. The EU New Life for Plastics Project set up a partnership in 1998 between local and authorities in north-west England and plastics recycling companies such as Reprise, Linpac, Plastic Packaging Services, Wellman and RECOUP to demonstrate sustainable plastic bottle recycling. The aim is to increase recycling plastic bottles from 2 per cent 25 per cent. Reprise says that the key to success in recycling is to work in partnership rather than in competition (Reprise, 1998, online).

Another example of a specialist environment damage limitation repair business is Shields Environmental Ltd. It is a medium-sized company employing 130 people, with a turnover of around £6 million, mainly concerned with dismantling and recovering electromechanical equipment such as redundant telephone exchanges. The company was set up in 1970 by Gordon Shields, whose aim was to explore the potential of recycling precious and rare metals. From the beginning the company set out to have stringent environmental standards. For a small and medium-sized company Shields Environmental has gone much further than many others. They have achieved a number of certifications such as EMAS, ISO 14001, ISO 9002, BS 775 and their environmental management system has been designed to be able to trace all of the hazardous substances from the point of collection through to final destination. The company points out that they go to such lengths because it makes good business sense, and gives them an edge over the competition (Business in the Environment, 1999, online).

The difficulty in classifying businesses unambiguously is shown by the case of remanufacturing. Remanufacturing, according to the Remanufacturing Industries Council International (RICRI), is where primary components from used products are dismantled, cleaned, restored, replaced, put in good order and then reassembled to work as if new. RICRI estimates the 73,000 remanufacturing firms have annual sales of $53 billion and employ 480,000 people in the USA. They operate in thirteen main sectors: motor vehicle parts, office furniture, compressors, electrical apparatus, vending machines, photocopiers, laser toner cartridges, data communications equipment, gaming machines, musical instruments, robots, aircraft parts and bakery equipment. The environmental advantages are savings on resource use and energy, as well as pollution and carbon emissions. And unlike recycling, where products are broken down into component elements, remanufacturing retains most of the original manufacture. Far less energy is required in remanufacturing, for example, 85 per cent of the energy used in the original automobile component manufacture is preserved in a remanufactured product. Is remanufacturing really an environmental damage limitation repair business? It has undoubted environmental benefits and clearly fulfils many objectives of such business but should it be properly included in the manufacturing sector? The example of Ford in setting up a remanufacturing business would suggest that it should be included there.

Environmental business services

We include in this section businesses whose purpose is to provide services relating to the environment. They include:

- the environmental media;
- environmental consultants;
- ethical investment trusts;
- environmental business clubs and associations;
- public/private partnerships.

The environmental media

We have already identified the media as a key player in environmentalism (Chapter 4). It provides news, information and opinion about environmental issues. It also influences events, alerts people, conducts campaigns and manipulates issues. It can be used by companies, pressure groups and politicians to advance their own agenda. Modern media is very diverse and serves different audiences. Table 9.3 indicates the principal types of media dealing with the environment.

Those that we term the 'environmentalist media' reach the people who already have more of an environmentalist viewpoint. They are a niche

Table 9.3 *Media and the environment*

Type of media	*Subtypes*
1 Environmentalist media	(a) specialist magazines, e.g. *The Ecologist, BBC Wildlife* (b) NPOs magazines, e.g. *Birds, Heritage Today* (c) specialist websites, e.g. Friends of the Earth, Greenpeace (d) environmental CD-ROMs (e) environmental videos, e.g. *Life on Earth, Raging Planet*
2 Media with substantial environmental coverage	(a) Magazines, e.g. *National Geographic, New Scientist* (b) Cable and satellite TV channels, e.g. *Discovery Channel*
3 General media with environmental content	(a) daily newspapers (varies considerably) (b) local newspapers (c) weekly and monthly general magazines (d) TV and radio news (e) TV and radio documentaries, e.g. *Land of the Tiger* (BBC TV) (f) feature films, e.g. *China Syndrome, Silent Running*
4 Trade media covering environment from business standpoint	(a) trade magazines, e.g. *Farmers Weekly, The Grocer* (b) business websites, e.g. *Business in Environment*
5 Academic media	(a) books (b) journals of academic societies (c) research reports (d) theses (e) university website

market and so are of interest to advertisers who can use it to market products such as books, videos, CD-ROMs, holidays and equipment.

Some media have substantial though not exclusively environmental coverage. *The National Geographic* is particularly significant because of its huge international readership. National newspapers and television cover the environment as part of their activities, in competition with every other issue. Some newspapers and channels are more concerned with the environment than others, e.g. in the UK the *Guardian* and Channel 4 are more likely to have substantial and serious coverage of the environment than the *Sun* or Channel 5.

Trade organizations are highly focused on the issues interesting particular industries, but they can be highly significant in promoting news and opinions. Some of them have a very large circulation, e.g. *Farmers Weekly* has over 200,000 readers. Finally, academic media provide the most impartial and detailed information and analyses and are often the basis for articles and programmes in the popular media. Academic media, have however, a much more restricted circulation and the more thorough analyses such as PhD theses are the least read. The explosive growth of the Internet has resulted in a huge platform for all sorts of publishing on the environment. Type in 'environment' on a web browser search box and a very large number of hits will be listed. A test on Yahoo turned up over 54 million hits for 'environment'. Assuming extreme dedication (and indeed longevity), it would take over 616 years just to read the home pages, at 2 minutes per website for 8 hours a day, 365 days a year!

Environmental consultants

Environmental consultancy is a growth industry. According to Howes *et al.* (1997, p. 6) environmental consultancy in the UK alone has become a £400 million per year business employing 12,000 people. The consultant is hired by business to carry out functions which it itself is unable or unwilling to perform because of lack of knowledge, expertise, time or confidence of success. It may be that the business itself does not have staff who are competent to address the issue, or even if it does, it hires a consultant to provide an impartial and credible viewpoint. An in-house environmental evaluation, for example, could run the risk of being blinkered by the corporate mind set and be unable to see solutions which are evident to an outsider. There is also the danger that an internal report might seek to provide the answers that it thought the organization wanted rather than the one they needed. A consultant therefore has the advantage not only of expertise, but also of objective independence. If the client does not like the report, then it need not act on it and at worst the consultant will not be hired again. However, an unwelcome internal report might be thought to have repercussions on the career prospects of the author and therefore lead to cautious or indeed useless results.

Consultancies undertake both general data collection and reporting functions for general publication in reports and websites, e.g. SustainAbility (1999, online), ENDS (1999, online) and EBI (1999, online) and client-specific activities such as surveys, reports, monitoring, audits, drawing up strategic plans and forecasting. Box 9.3 illustrates the activities of the EBI Consultancy. Consultancies can be complex organizations, SustainAbility, for example, says of itself:

> We are a bit of a hybrid organisation – we are described by others as a 'think tank', a 'campaigning organisation' and a 'consultancy'. All three are accurate up to a point, but our core business is consultancy. We are a for-profit limited company, although many of our goals are non-financial.
>
> (SustainAbility, 1999)

Box 9.3

EBI Consultancy

EBI – Environmental Business International Inc. is a US-based international environmental consultancy company operating since 1988 '[to] serve the business development needs of environmental companies by providing high quality strategic information products and services' (EBI, 1999). It collects and publishes information as well as providing specific services to clients. It publishes the *Environmental Business Journal*. Among the issues it consults over are: market research; competitive analysis; opportunity evaluation; strategic planning; telemarketing/lead generation; technology commercialization; new product development and new venture analysis.

Clients include government departments such as the Department of Commerce and the EPA, the OECD and a wide range of well-known companies such as DuPont, Hewlett Packard, Lockheed Martin, Mitsubishi International, Union Carbide and Westinghouse. The sort of services it offers include air pollution control, environmental information technology and software, sustainable development, waste management, recycling and water treatment management.

Source: EBI Consulting Services (1999)

They also make their living by providing information and recommendations to clients but they act to change the business climate generally through articles in the press and on their website. Without the development of environmental concern by business, they would not have developed in the way that they have.

Ethical investment and the environment

Historically there have often been tensions between the values of society and the practices of business. As Chandler and Wright (1999) point out, at one time businessmen vigorously defended their right to use slaves and child labour as essential to their profitability. Eventually they had to change because of a combination of social and political pressures. In the modern era, environment, human rights, social justice and health and safety have all become areas of ethical concern. As well as being a vital component of modern business, investment is an industry itself. Shareholder pressure is increasingly causing companies to take a more ethical and pro-active stance towards the environment.

Globalization has resulted not only in more world-wide business operations but also in more media coverage. The Internet has allowed angry individuals to achieve a platform they could never have reached in the past, and investigative reporters hunt for stories, the more scandalous the better. Multinational companies are very vulnerable to adverse publicity, e.g. Shell in Nigeria and Brent Spar, which can damage reputation, share values and sales. Hence the growing interest in ethical business practices. This receives added significance in the growth of ethical investment trusts which will only invest in businesses which meet a rigorous standard of behaviour.

Ethical investment trusts will invest in business involved in recycling, solar energy, water pollution control, organic farming, ecotourism, Third World producer co-operatives, etc. They will not invest in businesses involved in arms manufacture, tobacco, gambling, genetic engineering, factory farming or riot control equipment. Neither would they invest in companies which have a record of pollution, exploitation of employees or nature, support dictatorial regimes, contribute to poverty, social inequality or injustice. The essential point is that that they seek to make profits for their investors but not at the expense of people or nature. At one time this might have been regarded as hopelessly utopian and uncommercial – but then so was the idea of abolishing slavery in the eighteenth century.

The first ethical investment trust in the UK was Friends Provident 'Stewardship' set up in 1984 when few companies were concerned about the environment. Friends Provident was founded in 1832 by Quakers and had always taken an ethical outlook. By 1999 'Stewardship' had 100,000 investors with £1 billion funds under management (Friends Provident,

1999). The success of ethical funds is shown by the fact that there are now over 30 UK ethical funds with over £10 billion invested and £2 billion of that since 1996 (*WWF News*, 1999a). In the USA nearly 10 per cent of all investment funds are in ethical investments. The ethical funds are strict in selecting investments. Friends Provident say that 40 per cent of companies in the FTSE are acceptable for investment. This is clearly a powerful incentive for business and acts as an important influence. If business wants a share of these funds, then they need to be ethical not just to appear ethical. These developments have led to a growth of ethically validated audits and environmental reports. In the words of the Shell Social Audit team: 'It is not enough that companies behave responsibly, they have to be seen to be doing so' (quoted in Chandler and Wright, 1999, p. 26).

The Co-operative Bank has identified as stakeholders not only shareholders, customers and employees, but also local communities, national and international society and past and future generations (Green Futures, 1999). To this list could also be added suppliers as stakeholders. Ethical investing works because it delivers what SustainAbility (1999) calls the 'triple bottom line' or 'win-win-win business solutions which are socially responsible, environmentally sound and economically viable'. Increasingly, environmental ethics are becoming mixed with social justice and human rights. Environmental degradation frequently affects the poorest in society both at national and international levels. Poverty, ill health, social exclusion and repression are often bound up with polluted environments especially in developing countries.

Environmental business clubs and associations

Clubs and associations are an important way of diffusing best practice and raising standards among business communities. Businesses are more likely to pay attention to their peers who speak the same language. Business in the Environment (BiE) is a consortium of companies, environmental consultants, public environmental agencies and environmental think tanks which was established at the request of the Prince of Wales. Its aims are to promote the environment as a strategic business issue through practical guides and directories, training packs and research. Working to achieve sustainable development, BiE puts great emphasis on networking, putting companies in touch to learn from each other and disseminating best practice. Companies are grouped into regional and local clubs. Among the activities which clubs undertake are:

- conservation
- environmental audit
- environmental management systems
- finance
- help line
- meetings
- newsletters
- reviews
- training courses
- workshops
- written guidelines.

Most tailor their activities to the specific needs of their members, some groups run mock planning inquiries and court cases to familiarize members with issues of current concern. There are 109 business environmental clubs with over 11,500 members in total (BiE Club Directory, 1999, online). This is likely to be an underestimate since some clubs report membership as '200+' or 'N/A'. The regional distribution of clubs is shown in Figure 9.3 a and b. The Northwest has the greatest number of clubs and members, while Wales and Northern Ireland seem to have low representation. The figures show considerable geographical unevenness in club coverage. Since the purpose of the clubs is to network it is likely that the figures reveal a developing network rather than a completed one as the characteristic of spatial diffusion of organizations is

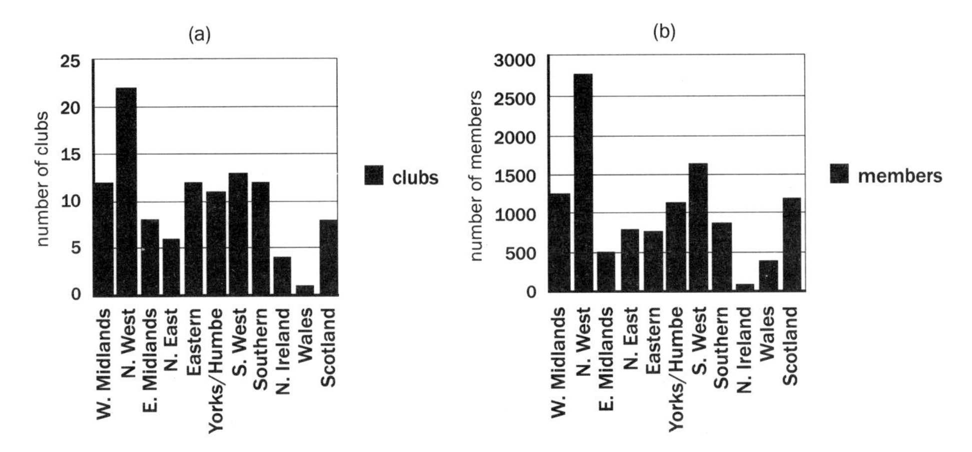

Figure 9.3 ***Environmental business clubs in the UK***
Source: BiE Club Directory (1999) online.

often geographical unevenness. They are an interesting development because they include SME businesses, which we have seen are less likely to have their own environmental policy compared to large enterprises. Sharing experience and emulating admired peers is an important way of diffusing best practice.

Another example is the US Environmental Bankers Association (EBA) which is a non-profit trade association of some sixty banking and financial institutions founded in 1994. The EBA is designed to do the following:

- to be a networking forum for exchange of environmental risk information;
- to act as a leading voice with government agencies;
- to provide information on all aspects of finance and the environment;
- to offer a technical resource for banking and professional associations.

The basic aim is to provide members with authoritative information and support when dealing with environmental matters, for example, how does a bank evaluate the risks of lending to a firm proposing to build a factory on contaminated land? The EBA has information and guidelines on how to approach such problems (EBA, 1999, online).

Trade associations such as the CBI and the IoD and NFU play an important role in influencing members' behaviour by endorsing or rejecting environmental legislation, lobbying legislators and promoting the interests of the trade generally. The interesting point here is that business has seen the need to set up specifically environmental clubs and associations in recent years.

Public/private partnerships

'Think globally; act locally' is a favourite slogan of the environmental movement. Cumulative small local actions on the environment, each in themselves insignificant, can add up to significant change overall. Almost every locality has some derelict or degraded land that would benefit the whole community if restored. It could add to global benefits in terms of carbon reduction if planted with trees, for example, but early attempts to deal with such problems which involved top-down action from public agencies seemed to be both expensive and limited in scope. The idea of partnerships to tackle environmental problems began in 1981 when the Groundwork Trust was set up to bring together business, the voluntary sector and local authorities as equal partners. Groundwork is a 'non hierarchical decentralised network' (Davidson, 1993) which works to shift attitudes as well as carry out tangible improvements. By 1997 there were 41 trusts covering 120 towns with a £25 million annual income of which business contributes 19 per cent (Jones, 1998). Each Groundwork Trust is a separate not-for-profit company and charity with a small core of professional staff who mobilize a whole host of other organizations. National Groundwork Headquarters co-ordinates and helps set up new projects, but it is the local trusts which are responsible for running the schemes on the ground.

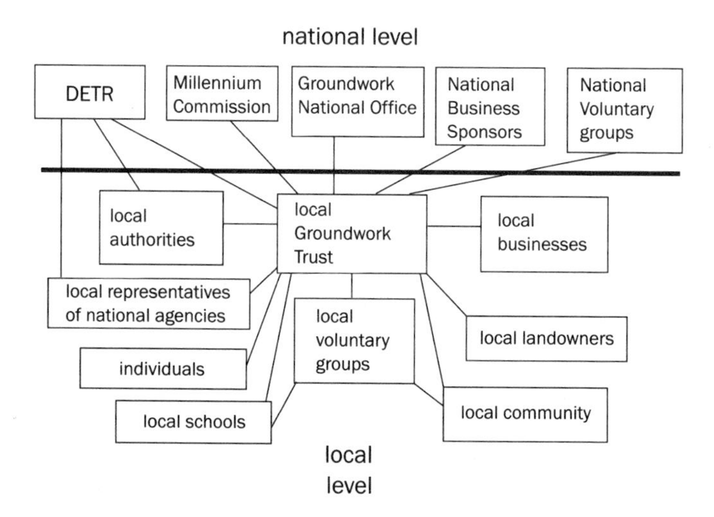

Figure 9.4 ***The Groundwork Trust network***
Source: From descriptions in Davidson (1993) and Jones (1998)

Figure 9.4 shows a typical Groundwork Trust network. Engaging the commitment of a wide range of bodies who might otherwise view each other with suspicion is a major challenge. Perhaps one of the most significant successes of Groundwork has been to convince SMEs to take part in environmental improvement projects. They have been persuaded to collaborate in reclaiming degraded land, transforming it into attractive business parks and have found that their business, employees and community relations improve. A more subtle process has also taken place with businesses being gradually persuaded to develop environmental policies and management systems and to become pro-active in looking for environmental opportunities. Groundwork has acted to create ecology parks in industrial estates with the aid of local voluntary groups and schoolchildren, which has not only benefited the environment, increased amenity and improved working conditions, but has also reduced vandalism. Other types of scheme involve the clean-up of polluted water courses, tree planting and making cycle paths on old railway lines. Working with BTCV, Groundwork contributes to practical training courses in conservation which can lead to permanent jobs in conservation. Groundwork therefore acts as a facilitator to get schemes off the ground. It also works with Environmental Business clubs and it is this that helps it to reach SMEs by focusing on the local and the achievable. Business becomes engaged because it can see benefits for itself as well as for the environment and the community.

A different kind of project, focusing on waste minimization which has environmental benefits is the Aire/Calder project begun in 1993 (Howes *et al.*, 1997). Eleven companies in West Yorkshire collaborated in a project with a number of public agencies to reduce their waste emissions to the local environment. The overt aim was to save costs for the companies as well as to help the environment. Various charitable trusts and public agencies provided 75 per cent of the financial cost of the project and the companies provided the remainder. A steering group representing the sponsors oversaw the project and consultants were appointed to manage it. A key ingredient was the generation of a culture of co-operation between the participants and for each company to have a project champion whose enthusiasm would act as a driving force. Senior management commitment was also vital for supporting the project and the underlying aim of pioneering procedures and techniques for further dissemination was also an important factor. Altogether, the companies between them identified 900 possible savings which saved £3.3 million a year. Monitoring, accurate data and training were important elements of the project and the success of the Aire/Calder scheme led to a further sixteen similar projects being set up. The lesson of partnership seems to be that business will take part in environmental improvement projects if it:

- is part of the decision-making team and not being dictated to;
- sees tangible business benefits;
- receives clear, accurate information on how to achieve objectives in ways it understands;
- works with organizations it trusts such as Business clubs;
- is made to feel part of the solution, not just the guilty culprit.

Summary

Despite the difficulties of classifying them, environmental businesses have been identified as falling into a number of classes. The first distinct group are conservation businesses which have developed as a result of the declining public funding and reorganization of public agencies, but at the same time the demand for conservation has grown. We have identified a new class called CANPOs which are the commercial arm of non-profit-making organizations. They employ a very wide range of business methods to raise funds for conservation. Other organizations are more concerned with making profits for business reasons but they need to be concerned about conservation since it provides the very reasons for their businesses. The example of the English Country House showed the very wide variations in approach. The second group of environmental businesses are the Environmental Damage Limitation and Repair Businesses (EDLRBs), who are more widely recognized as being 'green' businesses since their existence depends on cleaning up waste and pollution or installing and running systems which prevent environmental damage. Some of the functions of these firms can be internalized by large firms in other sectors, but the EDLRBs themselves do fill a distinct niche. The final group we identified is Environmental Services Business, which consists of the media, consultants, ethical investors, environmental business clubs and associations and public/private partnerships. They have all grown rapidly in recent years and are a sign of the increasing importance being accorded to the environment, both by the public and by business. Most of the environmental services identified are there to serve the needs of businesses. The fact that we have had to devise these classifications is a sign of the recent development of the sector and a sign of its growing sophistication. The reader should now be aware that the growing concern over the environment over the last two decades has led to a convergence between business and environmentalism that would have been unimaginable in the 1970s.

Further reading

A detailed examination of the role of conservation trusts is to be found in J. Dwyer and I. Hodge, *Countryside in Trust: Land Management by Conservation, Recreation and Amenity Organisation* (Chichester: Wiley, 1996). Two books which explain how countryside conservation is carried out are B. Green, *Countryside Conservation* (London: E and F Spon, 1996) and P. Bromley, *Countryside Management* (London; E and F Spon, 1990). Green looks more at management of habitats, while Bromley goes into a lot of practical details on things like visitor management and interpretation. P. Howes *et al.*, *Clean and Competitive?: Motivating Environmental Performance in Industry* (London: Earthscan, 1997) provides some very useful examples of the way industry has actually responded to the eco-modernist agenda. There are studies of public private partnerships and voluntary associations. M. Gandy, *Recycling and the Politics of Urban Waste* (London: Earthscan, 1994) considers the variety of ways waste is dealt with in big cities. The contrast between the municipal and various forms of private and public/private approach are examined in the context of a number of major cities such as London, New York and Hamburg. Finally for those who find the post-Thatcher system of public agencies, quangos, next steps agencies and public/private partnerships a bit confusing, M. Dynes and D. Walker, *The Times Guide to the New British State: The Government Machine in the 1990s* (London: Times Books, 1995), is an excellent aid.

Business and environment online: some websites

Much of the information for this chapter was researched on the Internet. The following represents some of the more useful websites concerned with environment and business. The addresses are current up to July 1999.

Business and the Environment
http://www.cutter.com/bate9811c.html

Business in the Environment
http://www.business-environment.org.uk/index2.html

Corporate Environmental Reporting
http://www.business-environment.org.uk/index2.html

ENDS-Environmental Data Services
http://www.ends.co.uk

Environmental Bankers Association
http://www.envirobank.org/

Environmental Business International Inc.
http://www.204.176.12.101/em...emID=I&componentID

Environmental Services Association
http://www.esauk.org/about_esa.html

Landmark Trust
http://www.travelrama.co.uk/44/landmark.html

Remanufacturing Industries Council International (RICI)
http://www.rici.org/frfaqust.htm

Reprise Ltd
http://www.ourworld.compuserve.com/homepage/repriseuk/about.html

Sustainable Business.com
http://www.sustainablebusiness.com

Transformation Strategies
http://www.trst.com/iso2a.htm

Discussion points

1 Why did conservation organizations feel it necessary to develop more commercial activities during the last two decades of the twentieth century?

2 Evaluate the relative benefits for a business joining a business-environment club and a Groundwork partnership.

Exercises

1 Find examples of each of the categories of conservation agencies and businesses shown in Figure 9.1. Search websites and publications to discover how far they use business methods in their conservation activities.

2 Conduct a recycling inventory of your own activities. What do you recycle? Why? If you don't recycle, why don't you? What could business do on a practical level to encourage you to recycle more? This could be expanded to a survey of colleagues or the general public. For a longer project, the scope could be expanded to finding out from local businesses what their recycling policies are.

3 Your company has been invited to join the local business-environment club. Check the Business in the Environment Club Directory to see what activities go on in your region (http://www.business-environment.org.uk/index2/html)

10 Environment and business: the future of the relationships

- The shape of the future – trends and patterns for the twenty-first century
- Future business impacts on the environment
- Future environmental impact on business
- Future impacts of environmentalism on business
- Future business impacts on environmentalism
- Conclusion

> I never think of the future. It comes soon enough.
>
> (Albert Einstein, 1930)

> Nobody is very good at forecasting what technologies people will want, invent, sell or buy.
>
> (von Weiszacker *et al.*, 1998)

> The trouble with our times is that the future is not what it used to be.
>
> (Paul Valéry)

Looking back at what people in the past predicted for the future can be an interesting and frequently amusing exercise, since even recent forecasts frequently turn out to have been dramatically wrong. The 1968 movie *2001: A Space Odyssey* predicted a future of moon bases and giant wheel-like space stations – none of which have come to fruition. Scientific prediction has little better a track record than science fiction. Brown and Flavin (1999) note that a group of distinguished futurists in 1890 were asked to make predictions for the next hundred years. They were universally optimistic about the technological and economic progress to come and foresaw a new utopia emerging. They correctly imagined that the increasing use of electricity, telephones, global free-trade and female emancipation would be extended. But they managed to miss out on predicting two world wars, nuclear weapons, the Internet,

Aids, global warming and the population explosion. It would of course have been much more surprising if they had, since there was no indication of any of those events in the 1890s.

The end of the twentieth century provoked a great deal of retrospective and prospective discussion in all fields. Speculation, reflection, prediction and wishful thinking were common. The field of environment and environmental business proved to be no exception (Von Weizsacker *et al.*, 1998; McLaren *et al.*, 1998; Brown and Flavin, 1999). Environmental forecasting has been on the business and political agenda since Meadows and Meadows *et al.*'s (1972) *Limits to Growth*, which galvanized the political and business establishments for the first time, not least because it used computer modelling and was thus 'scientific'. It therefore had an illusory credibility which the more emotional writings of many environmentalists did not possess. The central thesis that there are limits to economic growth in a finite world is undeniable, the room for argument is about when this will happen and what is necessary to prevent any eventual general economic and environmental collapse.

A host of environmental predictions and forecasts of the future have followed. Some have been deeply pessimistic such as Goldsmith, Hildyard *et al.* (1990) in *5000 Days to Save the Planet*. Some have extrapolated trends showing impending catastrophe without government intervention, for example, *Global 2000* (1982). Others have argued for the likelihood of the status quo being maintained (the so-called contrarian position) such as Simon and Khan (1984) and some have taken an optimistic view of a neo-cornucopian future. Von Weizsacker *et al.* (1998, p. 144) have described eco-capitalism as 'saving the earth for fun and profit through advanced resource efficiency'.

Change is certain but what the changes will be remain uncertain. While even short-term forecasts have a habit of coming unstuck, business and government do have to plan and they tend to make judgements based on projections. In most cases extrapolating past trends makes predictions of the future. The difficulty is that in the time it takes for investment decisions to be realized the world will have changed. There are a number of reasons why predicting the future frequently fails:

- The wrong questions may be posed due to cultural, corporate, social or political mind sets.
- The model used is incorrect. It does not incorporate all the causal factors.
- The data may be incomplete or inaccurate even if the model is right.

- The results may be misinterpreted.
- Future unpredictable changes not taken into account may invalidate the forecasts.
- The very act of forecasting alters the future by prompting counter-measures.

Despite all these problems, forecasts continue to be made. They have to be, since doing nothing is not an option. Nevertheless current methods of prediction do not have a very impressive track record. As von Weizsacker *et al.* (1998, p. 157) say (of economists' predictions, but it could apply to other fields), their pronouncements 'are only moderately more compelling and complete than earlier methods involving, say, inspecting the entrails of slaughtered fowl'.

Given all these problems, is there any merit in speculating on the future of environment and business? It does force a consideration about possible future directions of environment and business. We do not pretend to know what the future will be; all we can say is that there are certain futures, which may come about on the basis of present trends, and that businesses might respond in certain ways. The future business environment would change radically for example if the UK left the European Union, China became a multiparty democracy, commercial fusion power was developed, global warming was reversed by the onset of a new Ice Age, or a major earthquake hit Tokyo or Los Angeles. Bearing in mind these warnings we shall now turn to considering the possibilities.

The shape of the future – trends and patterns for the twenty-first century

The pace of change in the twentieth century has been unprecedented in all fields. The world's population grew fourfold in this century to six billion, global trade increased fifteen times, global wealth went up seven times, income per person went up six times and average life expectancy doubled (Brown and Flavin, 1999). This growth, however, was extremely unequal with 20 per cent of the world's population controlling 80 per cent of the wealth. The 'haves' are overwhelmingly in the developed industrial countries with the 'have nots' in the developing and underdeveloped countries. By the end of the twentieth century it was reported that the world's 225 richest individuals' wealth was equal to that of the three billion poorest and that the three richest multi-billionaires had wealth which exceeded the annual output of the 48 poorest countries (ibid.). Unprecedented economic growth has also led to unprecedented environmental stresses. What trends are likely to shape the future for business and environment? We identify three important areas to review:

- globalization;
- the New World Order;
- global environmental trends.

Globalization

Globalization (see also Chapter 1) is the idea that the world is becoming an integrated economic entity due to:

- world-wide instantaneous communication brought about by computers, satellites, fax, and the Internet;
- financial liberalization allowing a world-wide financial market and instant transfers of capital around the globe;
- liberalization of trade integrating all corners of the globe, while the creation of major trading blocs such as the EU and NAFTA accelerates the process of free trade within the blocs;
- multinational corporations shifting production plant to anywhere in the world to achieve the lowest costs;
- homogenization of culture to conform to a Western model, due to English-speaking, especially American, domination in the media, entertainment, sport and fashion;
- development of global consumer products for a world market, for example, Coca-Cola, Nike, McDonald's.

Virtually all governments have accepted as fact that globalization is inevitable and seek to embrace it for fear of being left out. However, as Massey (1999) points out, globalization is not some inevitable process but more an image of what powerful institutions such as the World Trade Organization want the world to become. It is a way of legitimizing the way development is taking place. Despite this, globalization is a very powerful force and has many important consequences for both business and environment. It tends to favour the bigger corporations who operate multinationally. Their financial power, R&D, and public relations resources can easily overwhelm local businesses, for example, the predicted impact of biotechnology firms patenting genetically modified seeds and thus giving them power over local food producers world-wide.

Globalization is often held to be responsible for many environmental effects such as the rapid deforestation of South-East Asia, where governments have sought to turn natural resources into cash, in order to develop, regardless of the environment consequences. This is likely to continue in the twenty-first century as it is too powerful a force to hold back.

The new world order

Compare any atlas of 1988, or earlier, with one produced today and the differences are stark. Borders have changed and new countries, or often revivals of old ones, have come into existence. Most notably the Soviet Empire has expired and the whole geopolitical basis of the bipolar Cold War world has gone. When the Cold War ended in 1990, there was much optimism about the new era of peace, with a peace dividend, allowing former weapons expenditure to be used for ending hunger, curing cancer, remedying environmental problems, etc. This mood of optimism lasted about six months until the Iraqi invasion of Kuwait in 1990. War and conflict had not ended with the Cold War and in fact seemed to take a nastier turn with vicious ethnic and religious conflicts in such places as Rwanda, Bosnia, Kurdish areas of Turkey and Iraq, Algeria, Sierra Leone, Liberia, Somalia and Afghanistan, Chechnya, Dagestan and Kosovo. Few people had even heard of most of these places a decade earlier.

The talk of a new world order following the end of the Cold War now seems premature. Some talk instead of a new world disorder and predict instead of stability, a spreading anarchy with countries being taken over by warlords, religious fanatics or organized crime (Kaplan, 1998). Most of the countries where these depressing events take place are the poor and marginal countries that are outside the limits of globalization at present. Such countries do not make good sites for investment. Companies are not likely to invest in countries which are disastrously unstable, dangerous to operate in and most of all, unprofitable because they lack much purchasing power. Almost their only value is for primary production. Will the twenty-first century see a rise or decline in these unstable areas? Will Russia surmount its current problems or will it sink into further decay? Until 1997 South Asian economies such as Malaysia and Indonesia were being heralded as a new miracle to match the NICs of South Korea, Taiwan, Singapore and Hong Kong. However, the rapid financial collapse of these countries, and the knock-on effect it had on all the others, showed the dangers of assuming that things will continue on present trends. What are the prospects for increased international regulations for environmental issues? Despite the plethora of international conferences, little has been done in practice to stem global warming. Nevertheless there are signs of increased international cooperation in the environmental field. For example, UNEP's International Declaration on Cleaner Production 1998 includes as signatories not only national governments, but also large corporations and international business organizations (Business and Environment, 1999, online).

Geopolitical predictions for the twenty-first century include wars in the Middle East with water replacing oil as the chief source of conflict. Further mismanagement of natural resources may lead to political turmoil if, for example, there were to be a repeat of the Indonesian forest fires of 1997 that spread smog throughout the region and weakened the tourist trade in other countries. There could be more local resistance to business, as in Ogoniland in Nigeria, where the local community and environment suffered from the effects of oil exploitation, but has not reaped the benefits.

Global environmental trends

The goal of almost all governments is still economic growth. As von Weizsacker *et al.* (1998, p. 153) put it: 'Economics is the state religion to which policies are aligned'. While countries in the developed world have become more environmentally aware and begun, albeit very slowly, to curtail the worst excesses of pollution and waste, developing countries lack the will, the resources, or the regulations to ameliorate environmental problems. With very few exceptions, the explicit goal of underdeveloped countries (UDCs) is to increase material standards of living through economic growth. This has involved the use of much dirty fuel technology, for example, most of India's and China's industrialization is based on coal. The race for development is increasing pollution so that cities in China now experience frequent smog. If everyone in the world aspired to US levels of car ownership, annual petroleum consumption would have to rise from 67 million barrels to 360 million barrels. If the world had American dietary standards, it would be necessary to cultivate land equivalent to four planets the size of the Earth

(Brown and Flavin, 1999, p. 15). Clearly, this is an impossible scenario. But developing countries are understandably annoyed at receiving lectures on environmental responsibility from countries that have already achieved higher living standards from plundering the Earth's resources and now want them to limit their growth for environmental reasons.

The global environmental trends for the next century are not very encouraging. Despite the rearguard disinformation of some contrarian supporters of fossil fuel companies, global warming is happening – fourteen of the warmest years since the mid-nineteenth century have occurred since 1980, and 1998 was the warmest year on record. Polar ice caps are melting at an accelerating rate and coral reefs are dying of heat stress. Extreme weather is becoming the norm world-wide with more frequent and more intense storms, hurricanes, floods and droughts. Insurance companies are paying out more each year for weather-related disasters and were one of the first industries to recognize the adverse impact of a warmer world. Insurance companies paid out $93 billion in 1990 for natural disasters. Since 1974 economic losses from natural disasters have risen nine times and insurance payouts have increased 15 times. The 1998 El Niño event was responsible for $14 billion worth of damage (Buckingham, 1999). The consequences of global warming include rising sea levels, which though slow, will have marked repercussions on low-lying areas as many of the world's major cities are coastal. They also involve impacts on food production, wildlife habitat, increased desertification and the spread of diseases. Threats to biodiversity seem to be increasing. The extinction of species has accelerated this century and shows no sign of decreasing. Some compare the scale of the current species extinction with that of the death of the dinosaurs 65 million years ago. No one can say precisely what the future environment will be, but it is clearly changing rapidly and while business has played a part in causing the problems, it also has the capability to both ameliorate conditions and to make commercial opportunities.

Future business impacts on the environment

Multinational companies are likely to continue their growth and because of their strength are best placed to anticipate trends and develop new responses. Both Shell and BP-Amoco have made considerable investments in non-fossil fuel energy development. Unless there is a nationalist backlash of high proportions, the likelihood is that globalization, free trade and financial liberalization will continue to spread their influence world-wide. We would thus expect to see an intensification of many present impacts, which include a draining of resources from South to North. The unfettered pursuit of free trade poses dangers to the environment. The World Trade Organization, in its desire to achieve a level playing field for trade, has declared that some countries' environmental regulations amount to unfair trade restriction. If a country says it will not import goods which do not meet certain environmental standards in their production, the WTO argues that this is discriminatory practice. If the WTO gets its way, environmental considerations would be forced down to the standards of the worst performers in the name of free trade. The outcome is at present uncertain.

A globalized world does not herald the 'end of geography' as predicted by O'Brien (1992). The fact that business can locate almost anywhere reinforces geographical differences and brings comparative advantage to the fore. Regional specialization will occur in those places where businesses judge they will thrive best, whether it be for reasons of natural resource endowment, cheap labour, skilled labour or favourable tax regimes. Technical innovation has long been a driving force of business and the twenty-first century is likely to see continuing advances in computer technology into all areas of life. For example, microchips are becoming so small, powerful and cheap that it will not be long before they are embedded in clothes in the form of 'smart fabrics' which could change colour, texture, or thermal properties according to programmed instruction. Already there exist prototype smart freezers that not only keep food at optimum temperatures, but also inform you via a screen on the door if the milk is running out and offer to order some more.

Von Weizsacker *et al.* (1998) chart fifty innovations that they say herald the advent of the new resource efficient economy. Substitution of renewable energy for fossil fuels, waste minimization, recycling and the adoption of integrated pollution control rather than 'end of pipe' clean-up technology are the main elements of ecological modernization. There is evidence that ecological modernization is happening in many advanced countries such as France, Germany, Denmark and the UK with lower resource consumption and lower pollution emissions, while still increasing economic growth. This has come about as a result of structural changes in their economies, what Gouldson and Murphy (1997, p. 77) call a shift from volume to value. Materials- and energy-intensive industries like steel-making have declined while those whose material and energy use is light such as retailing and tourism have increased. The irony of this is that it was not planned: 'It appears that advanced economies have begun to experience the effects that would be associated with a programme of ecological modernization without even trying' (Ibid., p. 77).

However, some of the reduction in resource use and pollution in developed countries is simply due to transferring the factories overseas to the newly industrialized countries. Less environmental impact in Europe perhaps, but more in south-east Asia. The environmental burden has simply shifted. As regulation and public attitudes harden towards pollution, this trend is like to continue. But on a global scale the pollution problem has not gone away.

The further penetration of the Internet is widely predicted to revolutionize the way we work and live (Green Futures, 1998). In a service-based economy with many workers spending most of their time word processing, reading and writing e-mail, finding information on the Internet and communicating with colleagues, bosses and clients, physical location ceases to matter. Provided one is wired up with the appropriate technology it doesn't matter if the worker is in Kensington, the Kyle of Lochalsh or Kuala Lumpur. An Internet economy could have profound environmental impacts, through a vast reduction in commuting, which would save huge amounts of energy and reduce pollution, let alone stress experienced by commuters.

Advances in biotechnology also promise changing future impacts. Genetically modified organisms have received a great deal of mostly unfavourable publicity in recent years, attracting the ire and direct action of environmentalists. They have a great potential for modifying the natural environment either to produce a cornucopia or to devastate biodiversity. At present we cannot be sure what might happen. Do bio-tech businesses have the potential to become the saviours or destroyers of the world? Hitherto unused or little used environments will become more used by business in the future. Further use of the ocean beds, Antarctica and near Earth orbital space is almost certain. Tourism for those who have been everywhere is poised to tackle the final frontier, with short flights into space, orbital hotels and eventually perhaps lunar holidays! How rapidly these things occur depends on the complex interaction between business cycles, technological development, government regulation and public demand.

Future environmental impact on business

How might environment affect business in the future? We consider some possible impacts of a number of issues.

Climate change

The changing climate is probably the biggest environmental issue that will affect business in the twenty-first century. Changing climates will have a strong influence on agricultural production as crop zones shift and farmers adapt, or fail to adapt, to changing conditions. At the macro level it could alter the balance of the world cereal trade if the American Midwest becomes drier and yields of corn and soya decline. At the micro level, the opportunities for extending vineyards northwards could lead to Scotland becoming as well known for its vintage wines as for its malt whiskies. Rising sea levels and the loss of coastal area will require relocation of business, or expensive flood protection. Warmer climates could see a boom for air conditioning companies but a decline for Harris Tweed. Undoubtedly there will be winners and losers but the smart businesses are those who will anticipate trends correctly.

Hazards and disasters

Partly as a result of changing climate, but also as a result of population expanding into geomorphologically and tectonically hazardous areas, the toll of natural disasters is likely to increase. Sudden and unexpected disasters can literally destroy businesses, but normally a business will be insured. However, the increasing frequency and severity of hazards are already leading to increased insurance premiums. Advances in GIS technology have allowed a much more precise evaluation of risks over space and thus marked differential insurance premiums. This could have an effect on influencing where businesses locate.

Urbanization

Urban areas, though a product of humanity, are increasingly important environments in their own right. The WRI (1998) estimate the world urban population to be 2.8 billion (47 per cent) in 2000 and predicts it to be 4.3 billion (57 per cent) in 2050. Of particular note is the growth of megacities, those above 8 million population. There were 2 in 1950, 23 in 1995 and by 2015 it is predicted there will be 36, of which 23 would be in Asia. The size of some of these megacities is now enormous, but they will be even bigger in the near future. Bombay is expected to grow from 15 million in 1995 to 26 million in 2015, Lagos from 10.2 million to 24.6 million, São Paulo from 16.5 to 20.3 million (Ibid., pp. 146–7). Urban growth on this scale could not only provide huge opportunities for house builders but also for civil engineers for infrastructure such as roads, airports, water, sanitation and power supply. Although much urban growth in the Third World cities at present is of self-build shanty type with appalling environmental conditions, it is possible that future urbanization will be of a higher standard which would provide huge opportunities for business.

Waste and pollution

In developed countries waste and pollution is being tackled with varying degrees of success. As we have seen in Chapter 9, environmental damage limitation and repair businesses have capitalized on the growing demand to curb or prevent pollution and waste. Manufacturing pollution control equipment is a growing industry stimulated by the demand for cleaner environmental conditions. As the developed countries have been growing relatively cleaner, the developing countries are becoming more polluted, for example, total suspended particulates (TSPs) in micro grams per cubic metre of air in Calcutta were 375, in Beijing, 366 and Lucknow, 463. By comparison the TSP count in Stockholm was nine, in Paris 40 and Athens 178 (WRI, 1998, table 8.5, pp. 64–5). Pollution levels in developing megacities are increasing and forecast to increase further. Eventually pollution control will become an issue in these cities and it will provide an opportunity for businesses to provide clean technology.

Biodiversity

In 1997 it was estimated that the value of the world's ecosystem services was $33.3 trillion compared to $18 trillion global GDP (Constanza, 1997, p. 256). Although these services, such as pollination, nutrient cycling and flood protection are free, if they did not exist in nature they would be a huge cost to provide artificially. Degradation of ecosystems and biodiversity has increased and despite protestations of governments that they are drawing up biodiversity plans, the assault on the natural world seems set to continue. Species of plants and animals can have economic value for use as medicine, fibre, oils, cosmetics and tourism, let alone their scientific and aesthetic value. Loss of species is not only a moral and cultural tragedy, it is also waste of potential economic benefits.

Resource depletion

Businesses which rely most on non-renewable resources such as coal and oil are ultimately doomed as in the end these resources will expire. However, given the short-run planning horizons of many businesses, the prospects of another thirty years of a resource might seem to be a very long time and offer little incentive to reduce consumption or to seek alternatives. The more far-sighted oil companies have already begun to diversify into renewable resources like solar and wind power. Even what should be a renewable resource such as forests or fish can be mined to extinction, though it is possible to replant trees and introduce new fish stocks. However, they may well not be the same varieties that have been destroyed. The challenge for twenty-first-century business is to adopt resource efficiency to make profits and help the environment.

Future impacts of environmentalism on business

The growth of environmentalism in the last three decades of the twentieth century was remarkable, not just through sheer expansion but also for its capacity to remain in the public eye, when other forms of organization such as trade unions and political parties declined in membership. Environmental politics has moved from the margin into the mainstream. Not only have environmental ministries become more significant, but Green Parties have had increasing success at the polls. In Europe Green Parties have long had national representation and for the first time in 1998 in Germany, a Green Party became a partner in the government coalition. Even in the UK where the voting system long marginalized minority parties, the Greens managed to get two MEPs elected in the 1999 European elections. If the lack of ideological difference between Left and Right continues, then it is likely the Green Party will attract more of the radical vote. Any increase in political power for the Greens is likely to result in further legislation and regulation of the environment, even if only because the conventional parties will seek to claw back lost votes by appealing to environmentalists.

Environmental pressure groups seem likely to at least maintain their momentum, especially if the environmental issues keep their prominence. Some pressure groups will develop further into quasi-businesslike organizations as they seek to fund their activities, while other non-land holding groups will give prominence to campaigning issues. Environmental media are likely to become ever more specialized and fragmented with the increasing development of the Web and digital television. This will enable advertisers to target market segments even more finely than the present. Environmental regulation is likely to increase, making environmental standards obligatory, which are at present voluntary. As regulation increases, so SMEs will become more obliged to follow the lead of large companies in drawing up environmental policies and management systems. Business behaviour is also likely to be affected by the strength of green consumer demand. If consumer demand increases, it would be suicidal for business to ignore it. How much will green consumerism increase and how fast? Although indications of present trends are that they will increase, there is much uncertainty. Is it possible, for instance, there could be a contrarian backlash or will demand reach a plateau, if environmental improvement reaches what most people regard as satisfactory, even if it is not truly sustainable? These are questions to ponder, but which are at present unanswerable.

Future business impacts on environmentalism

When the environment first became an issue in the 1960s implicating business as the culprit, the natural reaction of business people was to deny any such thing and hope that the problem would go away. Environmentalists were portrayed as cranks whose ideas were unproven and impractical. Ignore them and they will go away. When, on the contrary, environmentalists did not go away, but gained in strength and confidence to continue to challenge the business record on the environment, the next phase was to try to placate them through public relations by applying 'green cosmetics' or a 'green wash'. This made it look as if something were being done, while in reality changing nothing of substance. Such activities were soon exposed by groups such as

Friends of the Earth with their 'green con award' which gave businesses the wrong sort of publicity. During the 1980s and 1990s there seems to have been a genuine realization by some business leaders that the environment was a real issue of importance to them. Some experienced a 'road to Damascus'-style conversion like Ray Anderson of Interface (Armstrong, 1999) and began to work towards sustainability. As ecological modernization was developed and put increasingly into practice, business began to regard environmentalists in a new light – some as expert advisers who could help in their drive for eco-efficiency and others as a potential market. The process has been very uneven between sectors, with those most under fire, the energy, chemical and water industries taking action first. But it has spread to encompass service industries such as retailing and telecom groups. It is also most predominant among large corporations, with SMEs only recently becoming involved. The next century should see a filtering down to SMEs of the eco-efficiency idea while the bigger firms move into the next phase, that of sustainability.

Some like Welford (1997) warn of the danger of the 'corporate hijacking' of the environment. Business apparently embraces environmental issues, but only does so in order to set the agenda on its own terms. Business will only act on the environment if it feels it is in its own interest. It is unrealistic to suppose that businesses will act against their own interests. But we have moved away from the bipolar antagonism of 'either business or the environment'. The index of Corporate Environmental Engagement in Box 10.1, shows the extent to which leading companies have embraced the environment so far. As Brown and Flavin (1999, p. 19) say:

> The issue is not growth versus no growth, but what kind of growth and where. Converting the economy of the 20th century to one that is environmentally sustainable represents the greatest investment opportunity in history, one that dwarfs anything that has gone before.

If a sustainable economy is to be achieved, it will be through the creation of sustainable businesses. As Pretty (1998) shows, sustainability can be achieved in small steps and even if the business does not progress beyond a first step, at least it is going in the right direction and the chances are that it would be easier to take the next steps towards sustainability. One of the myths of converting to sustainability is that it will be costly and require tax rises. Roodman (1999), however, points out that at present 9 per cent of global government revenues ($650 billion) are spent in subsidies propping up unsustainable businesses, such as coal mining, fisheries and forestry. Cutting subsidies could reduce tax burdens by

Box 10.1

Corporate Environmental Engagement

The Index of Corporate Environmental Engagement was begun in 1996 by Business in the Environment and is an annual ranking of the FTSE 100 companies measuring what they are doing about the environment. It records whether or not a company has any of the following ten management practices:

- corporate policy on environment
- board member with responsibility for environment
- environmental targets
- environmental objectives
- stakeholder policy
- stewardship policy
- employee and environment policy
- environmental management system
- environmental audit
- environment and supplier policy.

In 1997, seventy-four companies took part and a further seven promised to take part next time. In addition, four ex-FTSE 100 continued to participate. Since the FTSE 100 is itself a league table, the membership changes from year to year. A significantly large proportion of the UK's biggest companies participate in the survey. As well as ranking the overall performance of companies, it is also possible to see which index measures are most or least adopted, e.g. over 80 per cent have a corporate environmental policy, but only 50 per cent have an EMS or Environmental Audit. The system of ranking provides a measure of competitiveness and due to the public availability of the results there is an incentive for companies to be seen to improve their performance. Considering the short time the index has been in existence, the level of participation is high; however, the index only measures the adoption of the various policies, it doesn't tell us how effectively they are employed.

Source: Business in the Environment: The Index of Corporate Environmental Engagement.
Available online: http://www.business-in-e...nment.org.uk/index2.html
Accessed 28/5/99

8 per cent. Taxes would shift to taxing things that were unwanted, like pollution, and away from things that were wanted, like jobs. Despite the logic of this, political reality often acts to modify, slow down or forestall such measures (Box 10.2). How to do the right thing may be obvious, but actually getting people to agree to do it is something else.

Box 10.2

Lobbying against the climate change levy

As part of its international obligations under the 1997 Kyoto Climate Conference, the UK Government planned a climate change levy or energy tax on industry with legislation due in the autumn of 1999 and a planned start date of 2001. The levy is based on European Commission suggestions which in turn are based on existing practices in a number of Member States, so it is not an entirely new idea. It is designed to encourage industry to be more energy-efficient, not simply to raise revenue as the money raised will be used to finance lower employer National Insurance contributions. However, even before the legislation was presented to Parliament, industry was lobbying against it in the summer of 1999. The main points of the controversy are:

1. Businesses from a wide range of industries cooperated to lobby against the proposed legislation in order to make it more amenable to their interests. These are tactics which have been successful in other cases (see Eden, 1996; Howes *et al.*, 1997).
2. Ministers are anxious to retain good relations with business for political reasons (New Labour worked hard to gain an image of business friendliness) and were likely to amend the legislation to make it more acceptable to business.
3. Industry would prefer tradable pollution permits so that individual businesses could decide if it were cheaper to buy a pollution permit or invest in energy saving.
4. The proposed levy was seen as a blunt instrument taxing all forms of energy and failed to discriminate between fossil and non-fossil fuels. It is argued that it would fulfil the aims of the tax better if it were the carbon content of the energy source which was taxed, thus encouraging, say, wind energy which has zero emissions at the expense of coal. However, this raises a further political complication as Labour has promised support for the remaining coalfield communities which form part of its core support.

Source: Crowe and Gow (1999).

Conclusion

In reviewing the past, present and future trends of environment and business we are struck by the rapidity of change in attitudes by businesses towards the environment in recent years. Business is becoming much more pro-active and has engaged in more constructive dialogue and partnerships with environmentalists. Governments now see themselves as facilitators and enablers more than playing a direct role as they might have done in the 1970s. Privatization and semi-privatization of government agencies have done much to encourage this process.

We wish to leave you with three sets of pointers to the future to consider:

- hopeful signs for the future;
- less hopeful signs;
- challenges for twenty-first-century environment and business.

Hopeful signs for the future

- Growth of increasing numbers of chief executive officers who have taken on board the environmental message and are putting it into practice. Leadership is vital especially in large multinational corporations.
- Growth of environmental audits. Corporate environmental reports have become an essential corporate accessory that leads to careful consideration of all aspects of the business in relation to the environment. It frequently leads to the realization that what is good for the environment is also good business practice.
- Growth of stricter environmental standards to live up to, for example, ISO 14000 and consequent peer pressure to conform to standards.
- Growth of shareholder pressure. Ethical Investment Trusts put pressure on businesses to act more environmentally responsibly if they want a share of this growing investment source.
- Growth of consumer awareness, growing demands for greener products and greener services. Increased environmental education and increased information availability mean that consumers know much more and will not fall for greenwash.
- Growth of environmental NGOs' membership, and growth of expertise in influencing business. NGOs have become more sophisticated in their campaigning.
- Growth of Green political parties. Growing significance of Green political parties means more likelihood of environmental legislation if only because mainstream parties seek to regain lost votes.
- Growth of international environmental regulation.
- The success of public–private partnerships in combining environmental and economic regeneration and forging links between businesses and environmentalists.
- Growth of environmental business clubs and associations which help to involve SMEs with environmental business.
- Growth of CANPOs to finance conservation objectives.
- Acceptance of sustainability as a goal by government and industry leaders.

Less hopeful signs

- Continued power of vested interests to keep 'business as usual'. Power still vested in the rich who defend their own interests though as the UK Green Party slogan during the 1999 European elections said: 'What's the point of being wealthy if you can't breathe the air?'.
- Greenwash hijacking of environmental agenda by business. Cosmetic approaches to environmental business problems can result in ineffective measures and disillusionment.
- Negative role of the World Bank and the World Trade Organization in promoting more unsustainable industrial development and blocking environmentally beneficial measures in the name of free trade.
- Globalization increasing pressure for profit at any price. This enables companies to move assets to countries with weak or non-existent environmental regulations to evade stricter environmental controls elsewhere. This is merely shifting environmental pollution to other locations without reducing the overall global environmental burden.
- Desperation of developing countries to catch up by using cheaper dirty technology, with the result that cities in China, Mexico and India have excessively high pollution levels. Primary producers unsustainably cutting down forests or pumping water tables dry in the name of development.
- International environmental legislation ineffective due to national interest blocs particularly by rich and powerful states who find ways to delay, water down or ignore agreements.
- Global warming continuing with little being done to stabilize it despite clear evidence of environmental deterioration such as die-off from thermal stress by coral reefs in the Indian Ocean, accelerated melting of the polar ice caps and increasingly unstable weather patterns leading to increased frequency of storms, droughts, floods and insect plagues.
- Growing disparity between rich and poor nations – poor people are not good customers therefore business is not interested in them.
- Increasing and continuing instability from war, ethnic conflict or crime. Collapse of civil governments undermining normal economic activity let alone sustainable development, for example in Kosovo, Bosnia, Iraq, Rwanda, Somalia, Sierra Leone, Liberia, Cambodia, Colombia and Kashmir.

Challenges for twenty-first-century environment and business

- Persuading more businesses that long-term sustainability is preferable to short-term unsustainable profits.
- Engineering changes in business culture so the environmental concerns are regarded as legitimate and necessary areas of business.
- Increasing adoption of eco-efficiency and ecological modernization because they are most likely to succeed with business. Business will only be engaged if it is beneficial to business to do so. Eco-radicalism is unlikely to succeed if it portrays itself as anti-business.
- Work towards the goal of the triple bottom line: profits, planet and people.
- Business working with government, NGOs, and consumers to achieve sustainability. A pluralist consensus approach that recognizes that no one section of the community can have it all its own way. Compromise is necessary to succeed.
- Reorientation of business and society to achieve goals rather than fixation on technologies or doing things just because they have always been done that way. For example, if the goal is to reduce carbon emissions, but at the same time society wishes to achieve the same levels of domestic heating or mobility, it must realize that doesn't have to be achieved by using fossil fuels or, by using internal combustion engines. There are alternative types of energy and technology that can achieve the same goals with less environmental damage.
- Work towards sustainability in small steps rather than try to convert all at once. Small steps are easier to take one at a time than leaping at breakneck speed three at a time and are easier to stop if a hazard looms ahead.
- Acceptance that there will be losers as well as winners in converting to sustainability, but this has always been so in every paradigm shift.
- Recognize that businesses that are pro-active and view the environment as an opportunity will fare better than those who resist or merely react to events.

Environment and business is an idea whose time has come. The explosion of environmental publishing in all forms in recent years demonstrates the growing interest in the subject. We hope that by deliberately taking a very broad and inclusive view of the subject, we have brought together concepts and examples that are not normally considered together in this way. This final chapter has been deliberately speculative and thus poses questions for which there are as yet no answers, but we hope it will stimulate the reader to think more about the subject critically and search for their own resolutions to the issues.

Further reading

For a useful guide to the changing geopolitical background: G. O'Tuathil, S. Dalby and P. Routledge, *The Geopolitics Reader* (London: Routledge, 1998) is recommended. It is one of the few political geography texts to include a substantial section on environmental geopolitics. Lester Brown and Christopher Flavin's *State of the World 1999 Special Millennium Edition* (London: Earthscan, 1999) is a Worldwatch report on progress towards a sustainable society, which is part of an annual series going back to 1994. This particular edition takes pains to review trends over the last century and considers how they will develop in the next. World Resources Institute, *World Resources 1998–99* (Oxford: Oxford University Press, 1998) is another annual publication which is an authoritative and accessible compendium of environmental statistics and comment. M. Jacobs (ed.), *Greening the Millennium? The New Politics of the Environment* (Oxford: Blackwell, 1997) has a lot more than the title would suggest which is of interest to students of environment and business. J. Elkington and J. Hailes, *Manual 2000: Life Choices for the Future You Want* (London: Hodder and Stoughton, 1998) is designed to provide the citizen with information about choices about almost everything from computers to funerals, cycling to addictions, greening the home to genetically modified food. Highly readable with a great many references, addresses and websites listed. D. McLaren, S. Bullock and N. Yousuf, *Tomorrow's World: Britain's Share in a Sustainable Future* (London: Earthscan, 1999) is a report for Friends of the Earth which shows in detail how the UK could become a sustainable society. An enthusiastic and optimistic book by E. von Weizsacker, A. Lovins and L. Lovins, is *Factor Four: Doubling Wealth, Halving Resource Use* (London: Earthscan, 1998). A very useful source of current business environmental activities, reviews and web sites is *Green Futures*, the magazine for Forum for the Future, a subscription-only publication available from: CIRCA, 13–17 Sturton Street, Cambridge CB1 2SN, or e-mail: greenfutures@circa-uk.demon.co.uk

Discussion points

1. Why is it difficult to forecast the future of environment and business with any degree of accuracy?
2. What grounds are there for optimism about future trends in business–environmentalism relationships?

3 What policies should governments adopt to ensure that business plays a positive role in environmental improvement?

Exercise

1 Examine past forecasts of the future for the year 2000 (such as *Global 2000* and *Resourceful Earth*). What did they predict? What actually happened? (Consult the latest statistics from the World Resources Institute.) Account for any discrepancies.

Glossary

acid rain form of precipitation containing dilute sulphuric acid. Acid rain is widely accepted as being formed by sulphur impurities derived from vehicle emissions and the burning of hydrocarbon fuels. The rain when it falls to earth causes damage to aquatic life, vegetation (particularly forests) and the external stone and metal surfaces of buildings.

affinity card a credit card which is operated by a financial institution but which bears the logo of a charity or group to which a percentage contribution of the balance is made. It provides a regular income to the charity, customers for the bank and the customer hardly notices the payments.

Agenda 21 arising from the 1992 Rio Conference this is a proposal that local communities should produce a plan for sustainability in their area. This was an advisory proposal which to date has been adopted by some environmentally minded local authorities world-wide.

Agriculture Act (1947) the cornerstone of post-war UK agriculture policy which encouraged modernization and growth of the farming sector. It promoted the development of a productionist ethic among farmers by subsidies, grants, incentives, tax breaks and advice. It differed from the CAP by operating a deficiency payment system, which paid farmers the difference between a notional agreed price and the world price, rather than putting up the real price for consumers.

Alkali Acts the series of UK pieces of legislation dating back to 1862 that forms the basis for the control of industrial air pollution.

biochemical demand (BOD) when pollutants, such as sewage, are put into rivers they will be cleaned up by bacteria in the river. These bacteria use up oxygen in the cleaning process. The amount of oxygen required to neutralize the pollution is known as the biochemical oxygen demand.

biodiversity the natural variety of life on Earth. Natural ecosystems tend towards complexity and diversity that are healthy and resilient. Loss of biodiversity through species extinction, habitat destruction, pollution and monocultures is regarded as a bad thing. Protecting and promoting biodiversity are the goals of environmentalists.

brownfield sites land which has already been used for industrial purposes and which is ripe for redevelopment. Properly applied to land that was previously used for industry and may now be derelict and/or polluted. It is increasingly used (in the UK) as a label for any potentially redevelopable land (such as former warehouses, old airfields, etc.) which is not greenfield.

Coalbrookdale area of the Severn valley at Ironbridge, Shropshire, where Abraham Darby produced the first marketable iron from a coke-based furnace in 1709. It is now regarded as being one of the cradles of the Industrial Revolution.

Common Agricultural Policy (CAP) the agricultural policy of the EU (and its predecessors). Designed to ensure a fair income for Europe's small farmers through a system of guaranteed prices, intervention buying, grants and supplementary payments in Less Favoured Areas (such as mountain regions), it became an engine for producing vast unsaleable surpluses (butter mountains, wine lakes) which enriched the largest agribusinesses and consumed the biggest proportion of the EU budget. Much criticized because of its high cost to taxpayers, consumers and the environment and feared because of the possible consequences on future enlargement to include the farmers of eastern and southern Europe, the CAP has become a target for reform. Some changes have been made to increase environmental incentives to farmers.

compliance the act of obeying regulations and laws. Companies comply with the regulations if they do not wish to be fined. It does not necessarily imply enthusiasm or willingness to obey.

contrarian label given to those who oppose environmentalism by arguing the opposite case, for example, a contrarian will deny that global warming is happening.

counter-urbanization the movement of urban people back to rural areas, not as farmers, but as commuters, retirees or telecommuters. Colonization of the countryside by urbanites.

Countryside Alliance an organization that coordinates the efforts of several rural pressure groups to lobby government on countryside issues. It came to public notice in 1998 when it organized the Countryside March. This involved 100,000 protestors demonstrating in London over a range of issues including the proposed ban on fox hunting, the 'beef on the bone' ban, low farm prices, housing expansion in the greenbelt, poor rural services and the proposed 'right to roam' legislation. Chiefly dominated by the pro-hunting lobby it managed to mobilize support from a wide range of rural interests with its slogan 'Listen to the Countryside'.

ecocentric branch of environmentalism that believes the way forward is through a sustainable economy in which natural ecological processes are allowed to flourish. This is usually considered to require zero economic and population growth with a strong community identity and a movement away from technological and capitalist solutions to environmental problems.

eco-efficiency the doctrine that environmental benefits can be obtained through business efficiency. It may be through introducing clean technology, recycling or energy efficiency, but the prime motive is for business to reduce costs as well as to cut pollution.

edaphic factors which relate to the soil forming processes as opposed to climatic influences

emissions trading a way of keeping pollution at agreed levels by putting a price on a given quantity of a pollutant, say, a tonne of carbon, and allowing producers to trade permits among themselves. One producer might introduce clean technology which allows them to reduce emissions. They can then sell the unused quota to another producer who either has not got the technology or cannot increase production without increasing pollution. In effect, it puts a price on pollution and turns it into a commodity that can be traded by business. The drawback is that it doesn't actually reduce the total pollution, it merely stabilizes it and shifts it around.

enclosure the process by which the communal open field system of medieval England was broken down into numerous small fields bounded by walls and hedges, each owned by a single landowner.

Enclosure allowed for new methods of animal husbandry and cropping. Although enclosure had begun in about the fifteenth century in some parts of England, it increased rapidly in the eighteenth century. The end of the eighteenth and the early nineteenth century saw a number of private Acts of Parliament completing the process.

European Union (EU) The group of fifteen European countries (at present) with joint economic and political policies (including environmental ones) which are supposed to ensure a common and harmonious approach. The membership and names of the organization have varied since its inception – Common Market, European Economic Community, European Community, European Union.

genetically modified organism a plant or animal which has been engineered by a biotechnology laboratory to contain certain genetic characteristics, e.g. pesticide resistance or increased yield. Vehemently opposed by environmental groups, genetically modified organisms have incurred market resistance initially in Europe and more recently in the USA.

globalization the process of economic and cultural homogenization of the world, brought about by the rapid advances of communications and information technology, free trade, multinational corporations, financial deregulation and spread of Western values and products. Depending on one's viewpoint, it is either the greatest blessing or the worst curse to affect the planet.

greenfield site land for development which has never been used for any urban or industrial activity and is currently open space such as agriculture, woodland, heath or moor. A greenfield site only comes into existence if a proposal is made for developing it; it is not a synonym for either the Greenbelt or the countryside in general. It has become regarded as the opposite of a brownfield, in the debate over the desirable balance between developing new sites (greenfields) and redeveloping old sites (brownfields).

greenwash environmental spin-doctoring. Deliberate distortion of information to make a company *appear* to be concerned about the environment while in reality doing little or nothing. A public relations exercise to make a firm look (unjustifiably) environmentally friendly.

ISO 14000 an international standard for environmental management systems audits, policies, etc. based on best practice. To achieve the standard, firms have to satisfy a range of criteria.

Laissez-faire the economic concept first advocated by Adam Smith that private enterprise and free trade will achieve greater economic growth than where governments intervene in the economic process.

limited liability company a company where the investors' liability in the event of the company failing is limited to the value of the share capital the investor has in the company.

Malthus, Thomas (1776–1834) economist and demographer famous for his belief that population growth would always tend to outrun food supply and that the limit to food supplies would always mean that without artificial control on population growth any natural surplus population would be checked by starvation and associated disease.

market segment a distinct population grouping with common purchasing behaviour. The object of market researchers' activities is to define ever more closely specific market segments on which to target advertising. Green consumers could be regarded as a market segment.

Minemata disease disease named after the town in Japan where a number of inhabitants developed a debilitating and life-threatening disease. The disease was ultimately traced to discharges of organic mercury from a local factory into the bay. The mercury entered the food chain since it was ingested by fish, which were subsequently eaten by the local population.

neo-cornucopian one who believes that technology and eco-efficiency will lead to an abundance of good things without the waste and pollution of the past.

Non-government organization NGO covers international organizations such as WTO, WHO, UNESCO as well as charities such as Oxfam and Greenpeace.

Non profit-making organizations NPO organizations with goals such as saving wildlife, e.g. WWF or conserving buildings, e.g. Civic Trust, but it does not mean that they do not engage in business activities such as catalogue merchandising or shops which may make a profit! The point is that any profit is ploughed back into the charity rather than going to shareholders.

Pigovian tax named after the early twentieth-century English economist, Arthur Pigou (1877–1959), who is widely regarded as a founding father of welfare economics. Such a tax would be introduced to achieve the optimal allocation of resources when a divergence becomes evident between private and social returns.

pluriactivity diversification of activities by farmers away from food production into farm tourism, conservation work, farm shops, craft work, forestry, etc. Pluriactivity offers a means of reducing over production of food in EU countries while retaining farmers on the land.

price support artificial raising of prices above free market or world prices by a national or supranational government in order to protect home producers from overseas competition. Particularly common with agricultural products in developed countries, but has been applied in the past to coal and some manufactured goods. Increasingly seen as anti-competitive in a global economy, it serves to protect the inefficient high-cost producer at the expense of the efficient low-cost foreign competitor.

quango variously defined as 'quasi-autonomous non-governmental organizations' or 'quasi-autonomous national government organizations', these are non-departmental official bodies dealing with a range of public functions that government prefers not to deal with directly, but feels disinclined to hand over completely to the private sector. Examples include English Heritage, English Nature, CADW.

retro fitting the technique of fitting equipment to industrial plant after the plant has been in operation for some time and when improvements to the current operating processes have been identified.

Rio Conference a conference held in Rio de Janeiro in 1992, attended by numerous high-level representatives of governments, NGOs and agencies to discuss the environment and development issues and to suggest practical means of implementing strategies for sustainability.

Romantics Nineteenth-century movement in art and literature, including the Wordsworths, whose prime concern was the preservation of God's glory but as the world got smaller felt the balance of nature was being disturbed. They saw no need to improve nature and resented any intrusion by urbanites into their rural beauty.

Spaceship Earth concept first used by Kenneth Boulding in his classic 1966 paper when he alluded to the Earth being likened to a spaceship. He regarded the Earth as a closed system in which matter cannot be produced or destroyed and so resources are not limitless and pollution and waste once created cannot be simply 'lost'.

stakeholder anybody who is considered to have a stake in an organization. This is taken to be wider than the owners of a company, extending to bankers, the local community, suppliers and subcontractors and perhaps even customers if their well-being is bound up with the fortunes of the organization.

steam coal variety of coal famously mined in South Wales and much prized for its high calorific value and low impurities that meant it gave off little smoke.

sustainability the Holy Grail of environmentalism. The idea of using renewable methods to ensure that there is no diminution of resources. If a tree is felled, a replacement should be planted. There are, however, two different approaches to sustainability. Environmentalists think business ought to be conducted in ways which sustain the environment; business thinks that sustainability is about using the environment (in an eco-efficient way) in order to sustain business. It is this difference which partly accounts for the upsurge in business interest in sustainability.

Taylorism pertaining to the views of Frederick Taylor, the early American management thinker, who developed the school of Scientific Management. These ideas measure all business processes and so led to the development of time and motion studies.

technocentric environmental tradition that maintains human ingenuity and technology will solve any outstanding problems. In any discussion of the environment and business this group is particularly important since much of the justification for business being green is that they are providing technological solutions to environmental problems.

transcendentalists nineteenth-century North American counterparts to the Romantics. The movement included personalities such as Ralph Waldo Emerson and John Muir. They saw the wilderness and particularly American wildlife being obliterated by the opening up of the continent under the influence of the rifle, the plough, barbed wire and the railway.

Utilitarians followers of the philosophical viewpoint of John Stuart Mill along with Jeremy Bentham and David Hume. They believed that actions should be judged in terms of the benefit they gave to society. They judged the value of inventions and developments in terms of society (rather than the private beneficiaries of developments) receiving the greatest well-being and happiness for the greatest number of people.

Utopians originally nineteenth-century group who believed in the wonders of nature and that the way forward was for mankind to live in harmony with nature. The proponents sought to prevent the destruction of communities through the onset of industrialization. This tradition has come down to the present day with the desire for sustainable futures and alternative technologies, shunning the use of 'big science'.

Websites

It may be that a number of these individual pages are no longer available. However, many of the sites do still exist and readers may find the sites useful in locating more recent relevant information posted.

Agriculture, Fisheries and Forestry Australia (AFFA) (1999) *National Landcare Programme*, Canberra: AFFA.
http://www.affa.gov.au/agfor/landcare/nlp.html accessed 7/1/2000.

British Wind Energy Association (1998).
http://www.bwea.com accessed 7/1/2000.

Business and the Environment (1998a) 'Recycling, remanufacturing in the spotlight', *Business and the Environment*, IX, 11, Nov. 1998.
http://www.cutter.com accessed 7/1/2000.

Business and the Environment (1998b) 'UNEP launches cleaner production declaration', *Business and the Environment*, IX, 11, Nov. 1998.
http://www.cutter.com/bate9811j.html accessed 30/6/1999.

Business in the Environment (1999a) 'Index of Corporate Environmental Engagement'.
http://www.business-environment.org.uk/index2.html accessed 28/6/1999.

Business in the Environment (1999b) 'Business clubs'.
http://www.business-environment.org.uk accessed 28/6/1999.

CMA (Society of Management Accountants of Canada) (1998) Publications Series Listings.
http://www.cma-canada.org accessed 7/1/2000.

Corporate Environmental Reporting website.
http://cei.sunderland.ac.uk/envrep/index.htm accessed 7/1/2000.

Day, R.M. (1998) 'Beyond eco-efficiency: sustainability as a driver for innovation', *Sustainable Enterprise Perspectives*, New York: World Resources Institute.
http://www.wri.org/wri/meb/sei/beyond.html accessed 7/1/2000.

Deloitte Touche Tohmatsu (1998) *Corporate Environmental Report Score Card: A Benchmarking Tool for Continual Improvement.*
http://www.teri-tohmatsu.jp/services/scorecard_E.html accessed 8/12/1998.

Department of the Environment Transport and the Regions (DETR) (1998) *Sustainable Business: Consultation Paper.*
http://www.environment.detr.gov.uk/sustainable/business/consult/opps.htm accessed 7/1/2000.

ENDS-Environmental Data Services.
http://www.ends.co.uk accessed 7/1/2000.

Environmental Bankers Association.
http://www.envirobank.org/ accessed 7/1/2000.

Environmental Business International Inc (1999) *EBI Consulting Services.*
http://www204.176.12.101/em...emID=I&componentID= accessed 7/1/2000.

Environmental Services Association (1998) *About ESA.*
http://www.esauk.org/about_esa.html accessed 7/1/2000.

European Parliament (1999) 'Reform of the CAP', *European Parliament Fact Sheet* 4.1.2.
http://www.europarl.eu.1.../factsheets/en/4_1_2.htm accessed 2/2/2000.

Friends of the Earth (1998) 'Green energy: which company should I choose?'.
http://www.foe.co.uk accessed 7/1/2000.

George, S. (2000) 'Fixing or nixing the WTO', *Le Monde Diplomatique*, January.
http: //www.corpwatch.org...feature/wto/6-george.html
accessed 02/02/00.

Greenpeace (1998) *News Release: Genetically Modified Organisms* 21/10/1998.
http://www.greenpeace.org.uk/press/pr00235.html accessed 21/10/1988.

Landmark Trust (1999) http://www.travelrama.co.uk/44/landmark.html. accessed 28/5/1999.

MAFF (1999) *Environmental Schemes.* http://www.maff.gov.uk/environ/envsch/default.htm accessed 7/1/2000.

Monsanto (1998a) *Biotech Primer: The Benefits of Biotechnology.* http://www.monsanto.co.uk/primer/benefits.html accessed 7/1/2000.

Remanufacturing Industries Council International (RICI) (1999) *Remanufacturing: Frequently Asked Questions.* http://www.rici.org/frfaqust.htm accessed 7/1/2000.

Reprise Ltd (1998) *Reprise.* http://www.ourworld.compuserve.com/homepage/repriseuk/about.html accessed 20/11/1998.

Ricoh (1999) '1999 environmental report'. http://www.ricoh-usa.com/about/environment/index.htm accessed 2/6/2000.

Sainsbury's (1997) *Interim Environmental Report.* http://www.j-sainsbury.co.uk/company/environment.htm accessed 15/1/1999.

Sainsbury, J. (1999) *Sainsbury's Environmental Report 1997/8*, London: J. Sainsbury Plc. http://www.j-sainsbury.co.uk/company/environment.htm accessed 7/1/2000.

SERA (2000) Socialist Environment Resources Association, 'We seek a greener Socialism', SERA Homepage. http://www.personal.u-net.com/~gaeia/sera.htm accessed 1/2/2000.

Shell (1998a) *Our Values, Principles, Commitments and Policy.* http://www.shell.com/values/0,1169,,00.html accessed 7/1/2000.

Shell (1998b) 'Protecting the planet'. http://www.shell.com/c/c2.html

Shiva, V. (1999) 'The historic significance of Seattle'. http://www.corpwatch.org...feature/wto/6-shiva.html accessed 2/2/2000.

SustainAbility Ltd (1999). http://www.sustainability.co.uk/ sustainability.htm.

Sustainable Business.com (1999) 'Chevron shareholders vote to track emissions', *Sustainable Business*.
http://www.sustainablebusiness.com accessed 18/5/1999.

Tilhill Economic Forestry (1998) *Forestry in the UK*.
http://www.tef.forestry.co.uk accessed 5/12/1998.

Transformation Strategies (1999a) ISO 14000 Concepts, *Transformation Strategies*.
http://www.trst.com/iso2a.htm
accessed 7/1/2000.

Transformation Strategies (1999b) ISO 14000 and environmental management case studies, *Transformation Strategies*.
http://www.trst.com/isocases.htm accessed 7/1/2000.

Bibliography

Adams, R. (1991) *Changing Corporate Values*, London: Kogan Page.

Adams, W.M., Bourn, N.A.D. and Hodge, I. (1992) 'Conservation in the wider countryside: SSSI's and wildlife habitats in eastern England', *Land Use Policy*, 9, 4, 235–48.

Allaby, M. (1986) *The Woodland Trust Book of British Woodlands*, Newton Abbott: David & Charles.

Allies, P. (1998) 'Don't buck the markets', *Green Futures*, 13, 7.

Andrews, J. and Rebane, M. (1994) *Farming and Wildlife: A Practical Management Handbook*, Sandy: RSPB.

Andrews, R.N.L. (1976) *Environmental Policy and Administrative Change*, Lexington, MA: Lexington Books.

Anink, D., Boonstra, C. and Mak, J. (1996) *Handbook of Sustainable Building*, London: James and James.

Arbury Hall (undated) *Arbury Hall: The Gothick Gem of the Midlands*, Nuneaton: Arbury Hall.

Armstrong, P. (1999) 'Ray's Hope', *Green Futures*, 14, 49.

Aston, M. (1985) *Interpreting the Landscape*, London: Batsford.

Bagwell, P. (1981) 'The decline of Isolation', in G.E. Mingay, *The Victorian Countryside*, London: Routledge and Kegan Paul.

Barney, G.O. (1982) *The Global 2000 Report to the President: Entering the 21st Century*, Harmondsworth: Penguin.

Bateman, D. and Ray, C. (1994) 'Farm pluriactivity and rural policy: some evidence from Wales', *Journal of Rural Studies*, 10, 1, 1–13.

Battarbee, R.W., Flower, R.J., Stevenson, A.C. and Rippey, B. (1997) 'Lake acidification in Galloway: a palaeoecological test of competing hypotheses,' in A.S. Goudie (ed.) *The Human Impact Reader*, Oxford: Blackwell.

Baumol, W. and Oates, W. (1988) *The Theory of Environmental Policy*, Cambridge: Cambridge University Press.

Bayliss-Smith, T. (1981) *The Ecology of Agriculture*, Cambridge: Cambridge University Press.

Beach, L.R. (ed.) (1998) *Image Theory: Theoretical and Empirical Foundations*, Mahwah, NJ: Lawrence Erlbaum Associates.

Beach, L.R. and Mitchell, T.R. (1987) 'Image theory: principles, goals and plans', *Acta Psychologica*, 66, 201–20.

Beaujeu-Garnier, J. and Delobez, A. (1979) *Geography of Marketing*, Harlow: Longman.

Beck, U. (1992) *Risk Society: Towards a New Modernity*, London: Sage.

Bell, C. and Newby, H. (1974) 'Capitalist farmers in the British class structure', *Sociologica Ruralis*, 14, 86–107.

Bell, M. (ed.) (1975) *Britain's National Parks*, Newton Abbott: David & Charles.

Bell, S. (1997) *Ball and Bell on Environmental Law*, 4th edition, London: Blackstone.

Benson, J. (1989) *British Coalminers in the Nineteenth Century*, London: Longman.

Benton, T. (1989) 'Marxism and natural limits', *New Left Review*, 178, 51–86.

Benton, T. (1993) *Natural Relations: Ecology, Animal Rights and Social Justice*, London: Verso.

Berry, G. and Beard, G. (1980) *A Century of Lake District Conservation*, Edinburgh: Bartholomew.

Best, R. (1981) *Land Use and Living Space*, London: Methuen.

Bishop, K.D. and Phillips, A.A.C. (1993) 'Seven steps to market – the development of the market-led approach to countryside conservation and recreation', *Journal of Rural Studies*, 9, 4, 313–38.

Blacksell, M. and Gilg, A. (1981) *The Countryside: Planning and Change*, London: George Allen and Unwin.

Blair, A.M. (1978) 'Spatial effects of urban influences on agriculture in Essex, 1960–1974', Unpublished PhD thesis, University of London.

Blair, A.M. (1980) 'Compulsory purchase', *Area*, 12, 183–90.

Blair, A.M. (1981a) 'Farming in Metroland', *The Countryman*, 85, 65–71.

Blair, A.M. (1981b) 'Farmers' attitudes to planning in the countryside', *Ecos: A Review of Conservation*, 2, 4, 6–11.

Blair, A.M. (1985) 'Farm-gate sales: direct retailing by farmers', *Grocery Business: Institute of Grocery Distribution Bulletin*, March, 15–20.

Blair, A.M. (1987) 'Future landscapes of the rural-urban fringe', in D. Lockhart and B. Ilbery (eds) *The Future of the British Rural Landscape*, Norwich: Geobooks.

Blaza, A. (1992) 'Environmental reporting – a view from the CBI', in D. Owen (ed.) *Green Reporting: Accountancy and the Challenge of the Nineties*, London: Chapman Hall.

Blowers, A. (1993) (ed.) *Planning for a Sustainable Environment*, London: Earthscan.

Blunden, J. and Curry, N. (eds) (1985) *The Changing Countryside*, London: Croom Helm.

Blunden, J. and Curry, N. (1988) *A Future for Our Countryside*, Oxford: Blackwell.

Blunden, J. and Turner, G. (1985) *Critical Countryside*, London: BBC.
Blythe, R. (1969) *Akenfield: Portrait of an English Village*, Harmondsworth: Penguin.
Body, R. (1982) *Agriculture: The Triumph and the Shame*, London: Temple Smith.
Body, R. (1984) *Farming in the Clouds*, London: Temple Smith.
Body, R. (1987) *Red or Green for Farmers (and the Rest of Us)*, Saffron Walden: Broad Leys.
Bonham-Carter, V. (1971) *The Survival of the English Countryside*, London: Hodder & Stoughton.
Bookchin, M. and Foreman, D. (1991) *Defending the Earth*, Montreal and New York: Black Rose Books.
Boulding, K.E. (1966) 'The economics of the coming Spaceship Earth', in H. Jarrett (ed.) *Environmental Quality in a Growing Economy*, Baltimore: Resources for the Future/Johns Hopkins Press.
Bower, J. and Cheshire, P. (1983) *Agriculture, the Countryside and Land Use*, London: Methuen.
Bowler, I.R. (1979) *Government and Agriculture: A Spatial Perspective*, Harlow: Longman.
Bowler, I.R. (1985) 'Some consequences of the industrialisation of agriculture in the European Community', in M.J. Healy and B.W. Ilbery (eds) *The Industrialisation of the Countryside*, Norwich: Geobooks.
Bowler, I.R. (ed.) (1992) *The Geography of Agriculture in Developed Market Economies*, Harlow: Longman.
Boyd, J.M. and Boyd, I.L. (1990) *The Hebrides: A Natural History*, London: Collins.
Bramwell, A. (1989) *Ecology in the 20th Century*, New Haven, CT: Yale University Press.
Braun, B. and Castree, N. (eds) (1998) *Remaking Reality: Nature at the Millennium*, London: Routledge.
Breheny, M.J. (1992) 'Towards sustainable development', in A. Mannion and S. Bowlby (eds) *Environmental Issues in the Nineties*, Chichester: Wiley.
Briggs, D. and Courtney, F. (1985) *Agriculture and Environment*, Harlow: Longman.
Bromley, D.W. (ed.) (1986) *Natural Resource Economics*, Boston: Kluwer-Nijhoff.
Bromley, P. (1990) *Countryside Management*, London: E & FN Spon.
Bromley, R. and Thomas, C. (eds) (1993) *Retail Change: Contemporary Issues*, London: UCL Press.
Brown, A. (ed.) (1992) *The UK Environment*, London: DOE/HMSO.
Brown, L. and Flavin, C. (1999) *State of the World 1999: A Worldwatch Institute Report on Progress Towards a Sustainable Society*, London: Earthscan.
Brown, P. (1998) 'Ecosoundings', *Guardian* 2 – Society, 14 October, p. 4.

Brown, P. (1999) 'Britain's all a twitter over birds', *Guardian*, 20 August.

Bryant, C. and Johnston, T. (1992) *Agriculture in the City's Countryside*, London: Belhaven.

Bryant, C., Russwurm, L. and McLellan, A. (1982) *The City's Countryside: Land and its Management in the Rural-Urban Fringe*, Harlow: Longman.

Buckingham, L. (1999) 'Claims for natural disasters rise to catastrophic high', *Guardian*, 16 March 1999, p. 19.

Budiansky, S. (1995) *Nature's Keepers: The New Science of Nature Management*, London: Weidenfeld and Nicolson.

Burgess, J. (1990) 'The production and consumption of environmental meanings in the mass media: a research agenda for the 1990s', *Transactions of the Institute of British Geographers*, NS15, 139–61.

Burgess, J. (1993) 'Representing nature: conservation and the mass media', in F. Goldsmith and A. Warren (eds) *Conservation in Progress*, Chichester: Wiley.

Burgess, J., Clark, J. and Harrison, C. (1998) 'Respondents' evaluations of a CV survey: a case study based on an economic valuation of the Wildlife Enhancement scheme, Pevensey Levels in East Sussex', *Area*, 30, 1, 12–17.

Burt, S. (1989) 'Trends and management issues in European retailing', *International Journal of Retailing*, 4, 4, 3–97.

Business and the Environment (1998a) 'Recycling, remanufacturing in the spotlight', *Business and the Environment*, IX, 11, Nov. 1998. Available online. http://www.cutter.com/bate9811c.html accessed 30/6/99.

Business and the Environment (1998b) 'UNEP launches cleaner production declaration', *Business and the Environment*, IX, 11, Nov. 1998. Available online. http://www.cutter.com/bate9811j.html accessed 30/6/99.

Buttel, F.H. (1986) 'Discussion: economic stagnation, scarcity, and changing commitments to distributional policies in environment-resource issues' in A. Schnaiberberg, N. Watts and K. Zimmerman (eds) *Distributional Conflicts in Economic-Resource Policy*. Aldershot: Gower.

CACI Ltd, (1994) *The Geodemographic Pocketbook: A Portrait of Britain's Products, Counties and Market Places*, Henley-on-Thames: CACI Ltd/NTC Publications

Cairncross, F. (1995) *Green Inc.: A Guide to Business and the Environment*, London: Earthscan.

Callan, S.J. and Thomas, J.M. (1996) *Environmental Economics and Management: Theory, Policy and Applications*, Chicago: Irwin.

Capital Shopping Centres plc (1998a) *Report and Accounts 1998*, London: CSC.

Capital Shopping Centres plc (1998b) 'Uxbridge CSC: the future', *A Great Shopping Partnership*, Issue 2, October 1998, London: CSC.

Capital Shopping Centres plc (1999a) 'A new name has been chosen for the new Uxbridge shopping centre; the Chimes', *A Great Shopping Partnership*, Issue 3, January 1999, London: CSC.

Capital Shopping Centres plc (1999b) *The Chimes Shopping Centre: Further information*, London: CSC.
Carlson, H. (1991) 'The role of the shopping centre in US retailing', *International Journal of Retail and Distribution Management*, 19, 6, 13–20.
Carrol, R. (1998) 'Why the farmers are angry. Again', *Guardian*, 28 May, p. 19.
Carruthers, S.P. (ed.) (1986) *Alternative Enterprises in UK Agriculture*, Reading: Centre for Agricultural Strategy.
Carson, R. (1962) *Silent Spring*, Harmondsworth: Penguin.
Castillon, A. (1992) *Conservation of Natural Resources. A Resource Management Approach*, Dubuque: Wm C. Brown.
Castree, N. (1995) 'The nature of produced nature: materiality and knowledge construction in Marxism', *Antipode*, 27, 12–48.
Cater, E. and Goodall, B. (1992) 'Must tourism destroy its resource base?' in A. Mannion and S. Bowlby (eds) *Environmental Issues in the 1990s*, Chichester: Wiley.
Chambers, J.D. and Mingay, G.E. (1966) *The Agricultural Revolution*, London: Batsford.
Champion, T. and Watkins, C. (1991) *People in the Countryside: Studies of Social Change in Rural Britain*, London: Paul Chapman.
Chandler, G. and Wright, M. (1999) 'Do the right thing – social accountability', *Green Futures*, 13, 22–8.
Chant, C. (ed.) (1989) *Science, Technology and Everyday Life 1870–1950*, London: Routledge.
Checkland, P. and Holwell, S. (1998) *Information, Systems and Information Systems*, Chichester: Wiley.
Checkland, S.G. (1982) *The Rise of Industrial Society in England 1815–1885*, Harlow: Longman.
Chisholm, M. (1979) *Rural Settlement and Land Use*, London: Hutchinson.
Clark, G. (1991) 'People working in farming: the changing nature of farmwork', in T. Champion and C. Watkins *People in the Countryside: Studies of Social Change in Rural Britain*, London: Paul Chapman.
Clarke, L. (1999) 'The places where we live', *Heritage Today*, 45, 34–40.
Clarke, P. (1999) 'Organic market is not exploited', *Farmers Weekly*, 29 January, p. 23.
Clarke, P., Jackman, B. and Merceer, D. (1980) *The Sunday Times Book of the Countryside*, London: Peerage Books.
Clayton, H. (1986) 'A sales drive without the hard sell', *The Times*, 23 April.
Coase, R. (1960) 'The problem of social cost', *Journal of Law and Economics*, 3, 1–44.
Coates, D.R. (1981) 'Subsurface influences', in K.J. Gregory and D.E. Walling (eds) *Man and Environmental Processes*, London: Butterworth.
Cobbett, W. (1830) *Rural Rides*, Harmondsworth: Penguin (reprint 1966 edition).
Cocker, M. (1998) 'Paradise lost', *Natural World*, 54, 20–3.

Constanza, R. (1997) 'The value of the world's ecosystem services and natural capital', *Nature*, 387, 252–64.

Cotgrove, S. (1976) 'Environmentalism and Utopia', *Sociological Review*, 24, 23–42.

Cotgrove, S. (1983) 'Environmentalism and Utopia', in T. O'Riordan and R.K. Turner *An Annotated Reader in Environmental Planning and Management*, Oxford: Pergamon.

Countryside Commission (1976) *The Lake District Upland Management Experiment*, Cheltenham: Countryside Commission.

Countryside Commission (1977) *Farm Open Days*, Cheltenham: Countryside Commission.

Countryside Commission/Victoria and Albert Museum (1986) *The Lake District: A Sort of National Property*, Cheltenham: Countryside Commission CCP194.

Coxall, W. and Robins, L. (1998) *Contemporary British Politics*. 3rd edition, Basingstoke: Macmillan.

Crabtree, J.R., Leat, P.M.K., Santarossa, J. and Thomson, K. (1994) 'The economic impact of wildlife sites in Scotland', *Journal of Rural Studies* 10, 1, 61–72.

Craig, D. (1990) *On the Crofters Trail: In Search of the Clearance Highlanders*, London: Jonathan Cape.

Crowe, R. and Gow, D. (1999) 'Labour's poll tax. Plans to charge a levy on industrial energy may bring an end to the government's current honeymoon with big business', *Guardian*, 29 August, p. 15.

Daniels, P.W. (1982) *Service Industries: Growth and Location*, Cambridge: Cambridge University Press.

Darling, F.F. (1947) *Natural History of the Highlands and Islands*, London: Collins.

Dartington Amenity Research Trust (1974) *Farm Recreation and Tourism in England and Wales*, Cheltenham: Countryside Commission CCP83.

Dartmoor National Park Authority (1989) *The Work of the Authority*, Bovey Tracey: DNPA.

Dartmoor National Park Authority (1995) *Annual Report*, Bovey Tracey: DNPA.

Davidson, J. (1993) 'Conservation and partnership: lessons from the Groundwork experience', in F. Goldsmith and A. Warren (eds) *Conservation in Progress*, Chichester: Wiley.

Davies, G.H. (1999) 'All credit to the wetlands', *Birds*, 17, 5, 43–6.

Davies, R.L. (1984) *Retail and Commercial Planning*, Beckenham: Croom Helm.

Dawson, J. (1979) *The Marketing Environment*, London: Croom Helm.

Dawson, J. (1983) *Shopping Centre Development*, Harlow: Longman.

Department of the Environment (1996) *Digest of Environmental Statistics*, No 18, London: HMSO.

Dicken, P. (1998) *Global Shift*. 3rd edition, London: Sage.

Dobson, A. (1990) *Green Political Thought*, London: Unwin Hyman.
Dodgshon, R.A. and Butlin, R.A. (1978) *An Historical Geography of England and Wales*, London: Academic Press.
Dorfman, R. and Dorfman, N.S. (eds) (1993) *Economics of the Environment: Selected Readings*, London: Norton.
Dryzek, J.S. (1997) *The Politics of the Earth: Environmental Discourses*, Oxford: Oxford University Press.
Dunn, J. (1998) 'Tapping into supply on demand. Two men are ready to take on the might of the privatised water companies', *Guardian*, 15 December, p. 21.
Dwyer, J.C. and Hodge, I.D. (1996) *Countryside in Trust: Land management by Conservation, Recreation and Amenity Organisations*, Chichester: Wiley.
Dynes, M. and Walker, D. (1995) *The Times Guide to the New British State; The Government Machine in the 1990's*, London: Times Books.
Eckersley, R. (1992) *Environmentalism and Political Theory: Toward an Ecocentric Approach*, London: UCL Press.
Eden, S. (1996) *Environmental Issues in Business*, Chichester: Wiley.
Edmond, H., Corcoran, K and Crabtree, B. (1993) 'Modelling locational access to markets for pluriactivity: a study in the Grampian Region of Scotland', *Journal of Rural Studies*, 9, 4, 339–50.
Edwards, A.M. and Wibberley, G.P. (1971) *An Agricultural Land Budget for Britain 1965–2000*, Studies in Rural Land Use Report No. 10, School of Rural Economics and Related Studies, Wye College.
Ehrlich, P.R. (1972) *The Population Bomb*, London: Pan/Ballantine.
Ehrlich, P.R. and Ehrlich, A. (1970) *Population, Resources, Environment: Issues in Human Ecology*, San Francisco: Freeman.
Ehrlich, P.R. and Ehrlich, A. (1987) *Earth*, London, Thames Methuen.
Einstein, A. (1930) quoted in *Oxford Dictionary of Quotations* (1992) p. 268, Oxford: Oxford University Press.
Elkington, J. and Burke, T. (1987) *The Green Capitalists: How to Make Money and Protect the Environment*, London: Victor Gollancz.
Elkington, J. and Hailes, J. (1988) *The Green Consumer Guide*, London: Gollancz.
Elkington, J. and Hailes, J. (1998) *Manual 2000: Life Choices for the Future You Want*, London: Hodder and Stoughton.
Elkington, J. and Knight, P. with Hailes, J. (1991) *The Green Business Guide*, London: Victor Gollancz.
Elsom, D. (1992) *Atmospheric Pollution: A Global Problem*, Oxford: Blackwell.
Elson, M.J. (1986) *Green Belts: Conflict Mediation in the Urban Fringe*, London: Heineman.
Encyclopeadia Britannica (1999) 'Brent Spar', Encylopedia Britannica Book of the Year 1996, Chicago: Encyclopedia Britannica Inc.
Engel, M. (1998) 'The day the city became a shire', *Guardian*, 2 March, p. 1.

English Heritage (1998) *The English Heritage Visitor's Handbook 1998–99*, London: English Heritage.
English Tourist Board/Rural Development Commission (1987) *A Study of Rural Tourism*, London: ETB.
Eurostat (1989) *Basic Statistics of the Community*, 26th edition, Brussels-Luxembourg: Statistical Office of the European Communities.
Evans, D. (1997) *A History of Nature Conservation in Britain*, London: Routledge.
Eyre, S.R. (1977) *The Real Wealth of Nations*, Oxford: Blackwell.
Fairbrother, N. (1970) *New Lives, New Landscapes*, London: The Architectural Press.
Fajer, E.D. and Bazzaz, F.A. (1992) 'Is carbon dioxide a "good" greenhouse gas?', *Global Environmental Change*, 294, 301–10.
Farmers Weekly (1998a) 'Rural Charter Launched', 29 May, p. 5.
Farmers Weekly (1998b) 'Close monitor will control GM crop release', 23 October, p. 6.
Farmers Weekly (1999a) 'Organic special', 12 February, 64–9.
Farmers Weekly (1999b) Editorial, 'Organic farming with realism could become godsend for industry', 12 February, p. 5.
Farmers Weekly (1999c) 'Precision farming '99', 12 March, p. 58.
Farm Holiday Bureau (1998) *Stay on a Farm 1998*, Norwich: Jarrold.
Fitzsimmon, M. and Goodman, M. (1998) 'Incorporating nature, environmental narratives and the reproduction of food', Chapter 9 in B. Braun and N. Castree (eds) *Remaking Reality: Nature at the Millennium*, London: Routledge.
Folmer, H., Gabel, H.L. and Opschoor, H. (1995) *Principles of Environmental and Resource Economics*, Aldershot: Edward Elgar.
Fotopoulis, T. (1997) *Towards an Inclusive Democracy*, London: Cassell.
Fox, W. (1984) 'Deep ecology: a new philosophy of our time?', *The Ecologist*, 14, 5/6.
Frankel, B. (1987) *The Post Industrial Utopians*, Cambridge: Polity Press.
Friends Provident (1999) 'Ethical investment: a thriving business', *Heritage Today*, 44, 14–15.
Fuller, R., Hill, D., and Tucker, G. (1991) 'Feeding the birds down on the farm; perspectives from Britain', *Ambio*, 20, 6, 232–7.
Gandy, M. (1993) *Recycling and Waste: An Exploration of Contemporary Environmental Policy*, Aldershot: Avebury.
Gandy. M. (1994) *Recycling and the Politics of Urban Waste*, London: Earthscan.
Garrett, A. (1998) 'From agent orange to tampered genes: Monsanto's life cycle', *The Observer*, 23 August, Business section, p. 3.
Garrod, G.D., Willis,K.G. and Saunder, C.M.(1994) 'The benefits and costs of the Somerset Levels and Moors ESA', *Journal of Rural Studies*, 10, 2, 131–46.

Gasson, R. (1973) 'Goals and values of farmers', *Journal of Agricultural Economics*, 24, 521–2.
Gershuny, J. and Miles, I. (1983) *The New Service Economy*, London: Pinter.
Giampietro, M., Cerretelli, D. and Pimental, D. (1992) 'Assessment of different agricultural production practices', *Ambio*, 21, 7: 451–9.
Gilg, A. (1978) *Countryside Planning: The First Three Decades*, Newton Abbott: David and Charles.
Gilg, A. (1996) *Countryside Planning*, 2nd edition, London: Routledge.
Giradet, H. (1992) *The Gaia Atlas of Cities: New Directions in Sustainable Urban Living*, London: Gaia Books.
Glasson, J., Therivel, R. and Chadwick, A. (1994) *Introduction to Environmental Impact Assessment*, London: UCL Press.
Glyptis, S. (1991) *Countryside Recreation*, Harlow: Longman.
Goldsmith, A. (1999) 'Villefranche-sur-Sainsbury's?', *Green Futures*, 14, 34–5.
Goldsmith, E. (1978) *Blueprint for Survival*, Harmondsworth: Penguin.
Goldsmith, E., Hildyard, N., Bunyard, P. and McCully, P. (1990) *5000 Days to Save the Planet*, London: Guild Publishing.
Goldsmith, F. and Warren, A. (eds) (1993) *Conservation in Progress*, Chichester: Wiley.
Gore, A. (1992) *Earth in the Balance: Forging a New Common Purpose*, London: Earthscan.
Goudie, A.S. (ed.) (1997) *The Human Impact Reader*, Oxford; Blackwell.
Goudie, A.S. (2000) *The Human Impact on the Natural Environment*, 5th edition, Oxford: Blackwell.
Gough, D. (1998) 'Big game hunters come to aid of quarry', *Guardian*, 12 November, p. 20.
Gouldson, A. and Murphy, J. (1997) 'Ecological modernisation: restructuring industrial economies', in M. Jacobs (ed.) *Greening the Millennium? The New Politics of the Environment*, Oxford: Blackwell.
Grant, W. (1993) *Business and Politics in Britain*, 2nd edition, Basingstoke: Macmillan.
Granville, M. (1998) 'Rural counter culture', *Green Futures*, 11, 46.
Green, B. (1996) *Countryside Conservation*, London: E. and F.N. Spon.
Green Futures (1998) 'The future of work in the virtual society', Special Report, 12, 24–39.
Green Futures (1999) 'The mushrooming of local produce', 16, 14.
Greenpeace (1998a) *Annual Review*, London: Greenpeace.
Gregory, D. (1978) 'The process of industrial change 1730–1900', in R.A. Dodgshon and R.A. Butlin *An Historical Geography of England and Wales*, London: Academic Press.
Grigg, D. (1989) *English Agriculture: An Historical Perspective*, Oxford: Blackwell.
Guardian (1999) 'Hurricane tours', 12 January, p. 12.

Guy, C. (1994) *The Retail Development Process: Location, Property and Planning*, London: Routledge.
Haigh, N. (1987) *EEC Environmental Policy and Britain*, 2nd edition, Harlow: Longman.
Haigh, N. (1993) *Manual of Environmental Policy: The EC and Britain*, Harlow: Longman.
Hall, P., Thomas, R., Gracey, H. and Drewett, R. (1973) *The Containment of Urban England*, 2 vols, London: George Allen and Unwin.
Hall, S. (1999) 'Royal Parks out of tune with concerts', *Guardian*, 27 April, p. 9.
Hall, T. (1981) *King Coal: Miners, Coal and Britain's Industrial Future*. Harmondsworth: Penguin.
Hanbury-Tennison, R. (1997) 'Life in the countryside', Countryside Alliance feature, *The Geographical*, LXIX, 11, 88–95.
Hanley, N., Shogren, J.F. and White, B. (1997) *Environmental Economics*, London: Macmillan.
Hanley, N. and Spash, C.L. (1993) *Cost Benefit Analysis and the Environment*, Aldershot: Edward Elgar.
Hannah, L. (1983) *The Rise of the Corporate Economy*, 2nd edition, London: Methuen.
Hapland, E. (1993) *Eco-renovation*, Dartington: Green Books.
Harbury, C. and Lipsey, R.G. (1993) *An Introduction to the UK Economy*, 4th edition, Oxford: Blackwell.
Hardin, G. (1968), 'The Tragedy of the Commons', *Science*, 162, 1243–8.
Harding, L. (1998a) 'Rural lobby raises its angry voice', *Guardian*, 2 March, 4–5.
Harding, L. (1998b) 'Rock'n'rural retreats' *Guardian*, 5 March, 6–7.
Hardy, A.R. and Stanley, P.I (1984) 'The impact of commercial agricultural use of organophosphorous and carbamate pesticides on British wildlife', in D. Jenkins (ed.) *Agriculture and the Environment*, Cambridge: Institute of Terrestrial Ecology.
Harris, F. and O'Brien, L. (1993) 'The Greening of shopping', in R. Bromley and C. Thomas (eds) *Retail Change: Contemporary Issues*, London: UCL Press.
Harrison, C.M. and Burgess, J. (1994) 'Social constructions of nature: a case study of the conflicts over Rainham Marshes SSSI', *Transactions of the Institute of British Geographers*, 19, 291–310.
Harrison, D. (1997) 'English Nature sparks peat row', *The Observer*, 30 November, 20.
Harvey, G. (1998) *The Killing of the Countryside*, London: Vintage.
Harvey, R.C. and Ashworth, A. (1993) *The Construction Industry of Great Britain*, Oxford: Butterworth-Heinemann.
Haughton, G. and Hunter, C. (1994) *Sustainable Cities*, London: Jessica Kingsley Publishers.

Hawken, P. (1993) *The Ecology of Commerce*, London: Weidenfeld and Nicolson.
Healy, M.J. and Ilbery, B.W. (eds) (1985) *The Industrialisation of the Countryside*, Norwich: Geobooks.
Henderson, H.J. (1993) *Paradigms in Progress*. London: Adamantine Press.
Hepworth, M. (1989) *Geography of the Information Economy*, London: Belhaven.
Hetherington, P. (1999) 'Shopping in surreal city under glass', *Guardian*, 11 March, p. 8.
Hill, B. (1993) 'The "myth" of the family farm: defining the family farm and assessing its importance in the European Community', *Journal of Rural Studies*, 9, 4, 359–70.
Hill, B. and Ray, D. (1987) *Economics for Agriculture: Food, Farming and the Rural Economy*, London: Macmillan.
Hill, H. (1980) *Freedom to Roam: The Struggle for Access to Britain's Moors and Mountains*, Ashbourne: Moorland Publishing.
HMSO (1990) *This Common Inheritance: Britain's Environmental Strategy* London: HMSO.
Hollis, G.E. (1997) 'Rain, roads, roofs and runoff: hydrology in cities', in A.S. Goudie (ed.) *The Human Impact Reader*, Oxford; Blackwell.
Hoskins, W.G. (1955) *The Making of the English Landscape*, London: Hodder and Stoughton.
Hough, M. (1995) *Cities and Natural Process*, London: Routledge.
Howard, E. and Davies, R. (1988) *Change in the Retail Environment*, Oxford Institute of Retail Management, Harlow: Longman.
Howes, P., Skea, J. and Whelan, B. (1997) *Clean and Competitive? Motivating Environmental Performance in Industry*, London: Earthscan.
Hunter, J. (1991) *The Claim of Crofting: The Scottish Highlands and Islands 1930–1990*, Edinburgh: Mainstream Publishing.
Hussein, A.M. (2000) *Principles of Environmental Economics*, London: Routledge.
Hutton, W. (1996) *The State We're In*, London: Vintage.
Ilbery, B.W. (1990) 'The adoption of the arable set-aside scheme in England', *Geography*, 75, 1, 69–73.
Ilbery, B.W. (1991) 'Uptake of the Farm Diversification Scheme in England', *Geography*, 76, 3, 259–63.
Ilbery, B.W. (1992) 'From Scott to ALURE – and back again?', *Land Use Policy*, 9, 2, 131–42.
Ilbery, B.W. (1998) *The Geography of Rural Change*, Harlow: Longman.
Ilbery, B.W. and Kidd J. (1992) 'Adoption of the Farm Woodland Scheme in England', *Geography*, 77, 4, 363–7.
Inglehart, R. (1977) *The Silent Revolution: Changing Values and Political Styles amongst Western Publics*, Princeton, NJ: Princeton University Press.

Institute of Business Ethics (1990) *Ethics, Environment and the Company: A Guide to Effective Action*, London: Institute of Business Ethics.
Institute of Environmental Health Officers (1991) *Environmental Health Report 1987–90*, London: IEHO.
Jacobs, M. (1991) *The Green Economy*, London: Pluto.
Jacobs, M. (ed.) (1997) *Greening the Millennium? The New Politics of the Environment*, Oxford: Blackwell.
January, D. (1999) 'To the ends of the earth', *Kew*, Richmond: Royal Botanic Gardens Kew, Spring, 24–9.
Jenkins, D. (ed.) (1984) *Agriculture and the Environment*, proceedings of ITE symposium No. 13, Cambridge: Institute of Terrestrial Ecology.
Jenkins, J. (1995) 'The roots of the National Trust', *History Today*, 45, 1, 3–9.
Jeremy, D.J. (1998) *A Business History of Britain, 1900–1990s*, Oxford: Oxford University Press.
Johnston, R. (1989) *Environmental Problems: Nature, Economy and the State*, London: Belhaven.
Jones, K. and Simmonds, J. (1990) *The Retail Environment*, London: Routledge.
Jones, P. (1998) 'The Groundwork Network', *Geography*, 83, 2, 189–96.
Kaplan, R. (1998) 'The coming anarchy', in G. O'Tuathail *et al.* (eds), *The Geopolitics Reader*, London: Routledge.
Kassler, P. and Patterson, M. (1998) *Energy Exporters and Climate Change*, London: Royal Institute of International Affairs.
Kennedy, M. (1998a) 'Films bring tourist boom for stately homes', *Guardian*, 26 February, p. 8.
Kennedy, M. (1998b) 'Stunning pastures new for farmers: attempts to cash in on Welsh scenic beauty', *Guardian*, 15 August, p. 15.
Kennedy, M. (1999) 'Heritage value is put on "pile of old stones"', *Guardian*, 20 January, p. 10.
Kerridge, E. (1967) *The Agricultural Revolution*, London: George Allen and Unwin.
Ketola, T. (1997) 'Ecological Eldorado: eliminating excess over ecology', in R. Welford *Hijacking Environmentalism: Corporate Responses to Sustainable Development*. London: Earthscan.
Kew (1999) 'Friendly companies', *Kew*, Royal Botanic Gardens, Kew, Spring, p. 45.
Kinnes, S. (1999), 'Organic fruit and vegetables. Shoparound Guide. *Guardian* 2, 14 January, p. 15.
Kirk, M. (1991) *Retailing and the Environment*, Harlow: Longman.
Klatte, E. (1991) 'Environmental and economic integration in the EEC', in Owen, L. (ed.) *Frontiers of Environmental Law*, London: Chancery.
Kohls, R. and Uhls, J. (1985) *Marketing of Agricultural Products*, New York: Collier Macmillan.

Krut, R. and Gleckman, H. (1998) *ISO 14001: A Missed Opportunity for Sustainable Global Industrial Development*, London: Earthscan.
Lake District National Park (1986) *Conserving the Lake District*, Kendal: LDNPA.
Lake District National Park Authority (1986) *Farming*, Kendal: LDNPA.
Larkham, P.J. (1996) *Conservation and the City*, London: Routledge.
Lawton, R. (1978) 'Population and Society (1730–1900)', in R.A. Dodgshon and R.A. Butlin, *An Historical Geography of England and Wales*, London: Academic Press.
Leach, G. (1979) *A Low Energy Strategy for the United Kingdom*, London: Science Reviews.
Leigel, L., Pliz, D. and Love, T. (1998a) 'The MAB Mushroom Study: background and concerns', *Ambio: A Journal of the Human Environment*, Special Report 9, September, 3–7.
Leigel, L., Pliz, D., Love, T. and Jones, E. (1998b) 'Integrating biological, socio-economic and managerial methods and results in the MAB Mushroom Study', *Ambio: A Journal of the Human Environment*, Special report 9, September, 26–33.
Leopold, L.B., Clarke, F.E., Hanshaw, B.B. and Balsley, J.R. (1971) *A Procedure for Evaluating Environmental Impact*, US Geological Survey, Circular no. 645, Washington, DC: US Geological Survey.
Lewis, M.N. (1992) *Green Delusions: An Environmentalist Critique of Radical Environmentalism*, Durham, NC: Duke University Press.
Lockhart, D.G. and Ilbery, B.W. (eds) (1987) *The Future of the British Rural Landscape*, Norwich: Geobooks.
Love, T., Jones, E. and Leigel, L. (1998) 'Valuing the temperate rainforest: wild mushrooming on the Olympic Peninsula Biosphere Reserve', *Ambio: A Journal of the Human Environment*, Special report 9, September, 16–25.
Lovelock, J. (1979) *Gaia*. Oxford: Oxford University Press.
Lowe, P. and Goyder, J. (1983) *Environmental Groups in Politics*, London: George Allen and Unwin.
Mabey, R. (1980) *The Common Ground: A Place for Nature in Britain's Future?*, London: Hutchinson.
MacAllister, D.M. (1980) *Evaluation in Environmental Planning*, Cambridge, MA: MIT Press.
McCarthy, J. (1998) 'Environmentalism, wise use and the nature of accumulation in the rural West', Chapter 6 in B. Braun and N. Castree (eds) *Remaking Reality: Nature at the Millennium*, London: Routledge.
MacEwan, A. and MacEwan, M. (1982) *National Parks: Conservation or Cosmetics?*, London: George Allen and Unwin.
MacEwan, A. and MacEwan, M. (1987) *Greenprints for the Countryside*, London: Allen and Unwin.
MacEwan, M. (ed.) (1976) *Future Landscapes*, London: Chatto and Windus.

McGoldrick, P. (1990) *Retail Marketing*, London: McGraw Hill.

McKibben, B. (1990) *The End of Nature*, Harmondsworth: Viking.

McKie, E. (1977) *The Megalith Builders*, London: Phaidon.

McLaren, D., Bullock, S. and Yousuf, N. (1998) *Tomorrow's World: Britain's Share in a Sustainable Future*, London: Earthscan.

MAFF (1989) *Planning Permission and the Farmer*, London: HMSO.

MAFF (1997) *UK Food and Farming in Figures*, Government Statistical Service MAFF, London: HMSO.

Mandler, P. (1997) *The Fall and Rise of the Stately Home*, New Haven, CT: Yale University Press.

Marren, P. (1993) 'The siege of the NCC: nature conservation in the eighties', in F. Goldsmith and A. Warren (eds) *Conservation in Progress*, Chichester: Wiley.

Massam, B.H. (1993) *The Right place: Shared Responsibilities and the Location of Public Facilities*, Harlow: Longman.

Massey, D. (1999) 'Imagining globalisation: power geometries of time-space', in Brah, A., Hickman, M. and Mac an Ghaill, M. (eds) *Global Futures: Migrations, Environment and Globalization*, London: Macmillan.

Mather, A. (1992) 'Land use, physical sustainability and conservation in Highland Scotland', *Land Use Policy* 9, 2, 99–110.

Mather, A. (1993) 'Protected areas in the periphery: conservation and controversy in Northern Scotland', *Journal of Rural Studies*, 9, 4, 371–96.

Maxwell, S. (1990) 'The rise of the environmental audit', *Accountancy* June 1990, 70–2.

Meadows, D., Meadows, D. and Randers, J. (1992) *Beyond the Limits: Global Collapse or a Sustainable Future*, London: Earthscan.

Meadows, D., Meadows, D., Randers, J. and Behrens III, W. (1972) *The Limits to Growth*, London: Pan and New York: Universe Books.

Mellanby, K. (1981) *Farming and Wildlife*, London: Collins New Naturalist.

Mercer, D. and Puttnam, D. (1988) *Rural England: Our Countryside at the Crossroads*, London: MacDonald.

Miller, F.A. and Tranter, R.B. (1988) *Public Perception of the Countryside* Reading: University of Reading, Centre for Agricultural Strategy.

Mills, L. and Schneegan, S. (1997) 'Greener local economic development: reflection from the UK on current approaches', *Local Environment* 2, 3, 231–47.

Mingay, G.E. (1976) *Rural Life in Victorian England*, London: Heinemann.

Mingay, G.E. (ed.) (1981) *The Victorian Countryside*, 2 vols, London: Routledge and Kegan Paul.

Mingay, G.E. (1990) *A Social History of the English Countryside*, London: Routledge.

Monbiot, G. (1998) 'Suffocate the golden goose: deregulation is just a subsidy for polluters and exploiters', *Guardian*, 5 November, p. 22.

Monsanto (1998) 'How does this potato differ from a potato?', Advertisement in *Guardian Weekend*, 15 August, p. 17.
Montagu, Lord (1967) *The Gilt and the Gingerbread: or How to Live in a Stately Home and Make Money*, London: Michael Joseph.
Montagu, Lord (1992) *40 Years at Beaulieu*, Beaulieu: Montagu Ventures Ltd.
Moore, E. (1991) 'Grocery distribution in the USA: recent changes and future prospects', *International Journal of Retail and Distribution Management* 19, 7, 18–24.
Moore, N.W. (1987) *The Bird of Time: The Science and Politics of Nature Conservation*, Cambridge: Cambridge University Press.
Morris, C. and Young, C.A. (1997) 'Towards environmentally beneficial farming? An evaluation of the Countryside Stewardship Scheme', *Geography*, 82, 4, 305–316.
Morris, H. and Romeril, M. (1986) 'Farm tourism in England's Peak National Park', *The Environmentalist*, 6, 2, 105–10.
Moss, G. (1981) *Britain's Wasting Acres: Land Use in a Changing Society* London: Architectural Press.
Muir , J. (1912) *The Yosemite*, first published New York: The Century Co., reprinted in *Eight Wilderness Discovery Books*, London: Diadem Books.
Munton, R. (1983) *London's Green Belt: Containment in Practice*, London: George Allen and Unwin.
Munton, R. and Collins, K. (1998) 'Government strategies for sustainable development', *Geography*, 83, 4, 346–57.
Murdoch, J. and Marsden, T. (1995) 'The spatialization of politics: local and national actor spaces in environmental conflict', *Transactions of the Institute of British Geographers*, 20, 3, 368–76.
Napier, T.L. and Sommers, D.G. (1994) 'Correlates of plant nutrient use among Ohio farmers: implications for water quality initiatives', *Journal of Rural Studies*, 10, 2, 159–71.
Newby, H. (1980) *Green and Pleasant Land?*, Harmondsworth: Penguin.
Newby, H. (1987) *Country Life: A Social History of Rural England*, London: Weidenfeld and Nicolson.
Newnham, D. (1999) 'Tomorrow's world', *Guardian Weekend*, 13 March, 38–9.
Nicholas, K. (1999) 'Faced with a clipboard consumers wax lyrical about the need to buy greener', *Green Futures*, 15, 54.
Nielsen Ltd (1994) *The Retail Pocketbook*, Henley-on-Thames: Nielsen/NTC Publications.
Nijkamp, P. and Perrels, A. (1994) *Sustainable Cities in Europe*, London: Earthscan.
North, J. and Spooner, D. (1982) 'Future for coal', *Town and Country Planning*, April, 93–98.
Norton-Taylor, R. (1982) *Whose Land is it Anyway?*, Wellingborough: Turnstone.

O'Brien, L. and Harris, F. (1991) *Retailing: Shopping, Society, Space*, London: David Fulton.

O'Brien, R. (1992) *Global Financial integration: The End of Geography*, London: Pinter.

O'Riordan, T. (1971) *Perspectives on Resource Management*, London: Pion.

O'Riordan, T. (1976) *Environmentalism*, London: Pion.

O'Riordan, T. and Turner, R.K. (1983) *An Annotated Reader in Environmental Planning and Management*, Oxford: Pergamon.

Orr, D. (1992) *Ecological Literacy: Education and Transition to a Post-Modern world*, Albany: State University of New York Press.

O'Tuathil, G., Dalby, S. and Routledge, P. (1998) *The Geopolitics Reader*, London: Routledge.

Owen, D. (ed.) (1991) *Green Reporting: Accountancy and the Challenge of the Nineties*, London: Chapman and Hall.

Owen, D. (1993) 'The emerging green agenda: a role for accounting', in D. Smith (ed.) *Business and the Environment*, London: Paul Chapman.

Owen, L. and Unwin, T. (eds) (1997) *Environmental Management: Readings and Case Studies*, Oxford: Blackwell.

Pallast, G. (1999) 'Fill your lungs – it's only borrowed grime', *The Observer, Business Section*, 24 January, p. 4.

Pawson, E. (1978) 'The framework of industrial change, 1730–1900' in R.A. Dodgshon and R.A. Butlin (eds) *An Historical Geography of England and Wales*, London: Academic Press.

Parry, M. (1990) *Climate Change and World Agriculture*, London: Earthscan.

Paul, R.W. (1988) *The Far West and the Great Plains in Transition 1859–1900*, New York: Harper Row.

Payne, J.W. (1976) 'Task complexity and contingent processing in decision making: an information search and protocol analysis', *Organizational Behavior and Human Performance*, 16, 366–87.

Pearce, D. (1983) 'Accounting for the future', in T. O'Riordan and R.K. Turner *An Annotated Reader in Environmental Planning and Management*, Oxford: Pergamon.

Pearce, D. Markandya, A. and Barbier, E. (1989) *Blueprint for a Green Economy*, London: Earthscan.

Pearce, D. and Turner, R.K. (1990) *Economics of Natural Resources and the Environment*, London: Harvester Wheatsheaf.

Pearce, G. (1994) 'Conservation as a component of urban regeneration', *Regional Studies*, 28, 1, 88–93.

Pearce, P.L. (1990) 'Farm tourism in New Zealand: a social situation analysis', *Annals of Tourism Research*, 17, 337–52.

Pepper, D. (1984) *The Roots of Modern Environmentalism*, Beckenham: Croom Helm.

Pepper, D. (1989/90) 'Green consumerism – Thatcherite environmentalism', *New Ground*, Winter, 18–20.

Perkin, H.J. (1981) *The Origins of Modern English Society*, London: Routledge and Kegan Paul.
Perkin, H.J. (1989) *The Rise of Professional Society*, London: Routledge.
Perman, R., Ma, Y. and McGilvray, J. (1996) *Natural Resource and Environmental Economics*, Harlow: Addison-Wesley Longman.
Phillips, D. and Williams, A. (1984) *Rural Britain: A Social Geography*, Oxford: Blackwell.
Piasecki, B.W., Fletcher, K.A. and Mendelson, F.J. (1995) *Environmental Management and Business Strategy*, Chichester: Wiley.
Pliz, D., Brodie, F., Alexander, S. and Molina, R. (1998) 'Relative value of Chanterelles and timber as commercial forest products', *Ambio: A Journal of the Human Environment*, Special report 9, September, 14–16.
Pocock, D. (1997) 'The UK World Heritage', *Geography*, 82, 4, 380–5.
Pollard, S. (1992) *The Development of the British Economy 1914–1990*, London: Edward Arnold.
Ponting, C. (1991) *A Green History of the World*, London: Sinclair Stevenson.
Porchester, Lord (1977) *A Study of Exmoor*, London: HMSO.
Porritt, J. (1984) *Seeing Green: The Politics of Ecology Explained*, Oxford: Blackwell.
Porritt, J. (1990) *Where on Earth are we Going?*, London: BBC.
Porritt, J. (1991) *Save the Earth*, London: Dorling Kindersley.
Porritt, J. (1997) 'Environmental politics: the old and the new', in M. Jacobs (ed.) *Greening the Millennium? The New Politics of the Environment*, Oxford: Blackwell.
Pretty, J. (1998) *The Living Land: Agriculture, Food and Community Regeneration in Rural Europe*, London: Earthscan.
Price, D.G. and Blair, A.M. (1989) *The Changing Geography of the Service Sector*, London: Belhaven.
Punch, M. (1996) *Dirty Business*, London: Sage.
Pye-Smith, C. and North, R. (1984) *Working the Land*, London: Temple Smith.
Rackham, O. (1986) *The History of the Countryside*, London: Dent.
Read, A, Phillips, P. and Robinson, G. (1998) 'Professional opinions on the evolving nature of the municipal solid waste management industry', *Geography*, 83, 4: 331–45.
Reed, M. (1990) *The Landscape of Britain from the Beginnings to 1914*, London: Routledge.
Rees, W. (1997) 'Is "sustainable city" an oxymoron?', *Local Environment*, 2, 3, 303–10.
Reynolds, J. (1993) 'The proliferation of the planned shopping centre', in R. Bromley and C. Thomas (eds) *Retail Change: Contemporary Issues*, London: UCL Press.
Rhind, D. and Hudson, R. (1980) *Land Use*, London: Methuen University Paperbacks.

Robinson, G. (1990) *Conflict and Change in the Countryside*, London: Belhaven.
Rogers, R. (1991) 'An overview of American retail trends', *International Journal of Retail and Distribution Management*, 19, 6, 3–12.
Roodman, D. (1999) 'Building a sustainable society', Chapter 10 in L. Brown and C. Flavin *State of the World 1999: A Worldwatch Institute Report on Progress Towards a Sustainable Society*, London: Earthscan.
Rose, C. (1990) *The Dirty Man of Europe: The Great British Pollution Scandal*, London: Simon and Schuster.
Rowell, A. (1996) *Green Backlash: Global Subversion of the Environmental Movement*, London: Routledge.
Ryle, M. (1988) *Ecology and Socialism*, London: Century Hutchinson.
Sainsbury's (1998a) *The Living Landscape: Our Commitment to a Sustainable Countryside*, London: Sainsbury's Supermarkets Ltd.
Sainsbury's (1998b) *Fresh Idea: Your Nearest Sainsbury's Home and Away*, London: Sainsbury's Supermarkets Ltd.
Sainsbury's (1998c) *Genetically Modified Soya*, London: Sainsbury's Supermarkets Ltd.
Sainsbury's (1999a) *Sainsbury's Environmental Report*, London: J. Sainsbury plc.
Sainsbury's (1999b) *Sainsbury's Organic Foods*, London: Sainsbury's Supermarkets Ltd.
Sandbach, F. (1980) *Environment, Ideology and Policy*, Oxford: Blackwell.
Sandbach, F. (1982) *Principles of Pollution Control*, Harlow: Longman.
Schama, S. (1995) *Landscape and Memory*, London: HarperCollins.
Schumacher, E. (1989) *Small is Beautiful: Economics as if People Really Mattered*, New York: Harper & Row.
Seddon, Q. (1989) *The Silent Revolution*, London: BBC.
Seymour, J. and Giradet, H. (1987) *Blueprint for a Green Planet*, London: Dorling Kindersley.
Sheail, J. (1974) 'The legacy of historical times', in A. Warren and F. Goldsmith (eds) *Conservation in Practice*, Chichester: Wiley.
Sheail, J. (1976) *Nature in Trust: The History of Nature Conservation in Britain*, Glasgow: Blackie.
Shoard, M. (1982) *Theft of the Countryside*, London: Temple Smith.
Shoard, M. (1987) *This Land is Our Land*, London: Paladin.
Simme, J., Olsberg, S. and Tunnell, C. (1992) 'Urban containment and land use planning', *Land Use Policy*, 9, 1, 36–46.
Simmons, I.G. (1996) *Changing the Face of the Earth: Culture, Environment, History*, 2nd edition, Oxford: Blackwell.
Simms, J. (1992) 'Green issues and strategic management in the grocery retail sector', *International Journal of Retail and Distribution Management*, 20, 1, 32–42.

Simon, J. and Kahn, H. (eds) (1984) *The Resourceful Earth: A Response to Global 2000*, Oxford: Blackwell.
Sinclair, D. (1990) *Shades of Green: Myth and Muddle in the Countryside*, London: Grafton.
Slee, B. (1989) *Alternative Farm Enterprises*, Ipswich: Farming Press.
Smales, J. (1999) 'Pleasure gardens and productivity', *Green Futures*, 14, 37.
Smith, D. (ed.) (1993a) *Business and the Environment: Implications of the New Environmentalism*, London: Paul Chapman.
Smith, D. (1993b) 'Business and the environment: towards a paradigm shift', in D. Smith (ed.) *Business and the Environment*, London: Paul Chapman.
Smith, M., Whitelegg, J. and Williams, N. (1998) *Greening the Built Environment*, London: Earthscan.
Stamp, L.D. (1969) *Man and the Land*, London: Collins.
Street, A.G. (1932) *Farmer's Glory*, Oxford: Oxford University Press, 1983 reprint.
Symes, D. and Marsden, T. (1985) 'Industrialisation of agriculture: intensive livestock farming in Humberside', in M.J. Healy and B.W. Ilbery (eds) *The Industrialisation of the Countryside*, Norwich: Geobooks.
Tarrant, J. (1984) *Agricultural Geography*, Newton Abbott: David and Charles.
Taylor, S.R. (1992) 'Green management: the next competitive weapon', *Futures*, September, 669–80.
Thomas, C. and Bromley, R. (1998) 'The impact of out of centre retailing', in R. Bromley and C. Thomas (eds) *Retail Change: Contemporary Issues*, London: UCL Press.
Thomas, D. (1970) *London's Green Belt*, London: Faber and Faber.
Thomas, J. (1999) 'Fill up and cut pollution', *Green Futures*, 13, 41.
Thompson, S. and Therivel, R. (eds) (1991) *Environmental Auditing*, Working Paper No. 130, School of Biological and Molecular Sciences and Planning, Oxford: Oxford Polytechnic.
Tickell, O. (1998) 'LETS on a leash', *Green Futures*, 12, 44–5.
Tickell, O. (1999) *The Earth Centre, A Geographical Special Supplement*, June.
Tietenberg, T.H. (2000) *Environmental and Natural Resource Economics*, 5th edition, Reading, MA: Addison-Wesley Longman.
Townsend, A. (1991) 'New forms of employment in rural areas: a national perspective', in T. Champion and C. Watkins *People in the Countryside: Studies of Social Change in Rural Britain*, London: Paul Chapman.
Townsend, A. (1993) 'The urban–rural cycle in the Thatcher growth years', *Transactions of the Institute of British Geographers*, 18, 2, 207–21.
Treadgold, A. (1990) 'The developing internationalisation of retailing', *International Journal of Retail and Distribution Management*, 18, 2, 4–11.
Turner, R.K. (ed.) (1993) *Sustainable Environmental Economics and Management*, London: Belhaven.
Turner, R.K., Pearce, D. and Bateman, I. (1994) *Environmental Economics*, London: Harvester Wheatsheaf.

Uzzell, D. (ed.) (1989a) *Heritage Interpretation*, vol 1, *Natural and Built Environment*, London: Belhaven.

Uzzell, D. (ed.) (1989b) *Heritage Interpretation*, vol 2, *The Visitor Experience*, London: Belhaven.

Walker, H.J. (1997) 'Man's impact on shorelines and nearshore environments', in A.S. Goudie (ed.) *The Human Impact Reader*, Oxford; Blackwell.

Wallace, D. (1995) *Environmental Policy and Industrial Innovation Strategies in Europe, the US and Japan*, London: Royal Institute of International Affairs.

Walthern, P. (ed.) (1988) *Environmental Impact Assessment*. London: Unwin and Hyman.

Ward, N. and Lowe, P. (1994) 'Shifting values in agriculture: the family farm and pollution protection regulation', *Journal of Rural Studies*, 10, 2, 173–84.

Ward, S. (1991) *The Countryside Remembered*, London: Century.

Warren, A. and Goldsmith, F. (eds) (1974) *Conservation in Practice*, Chichester: Wiley.

Warren, A. and Goldsmith, F. (eds) (1983) *Conservation in Perspective*, Chichester: Wiley.

Warren, S. (1995) *From Margin to Mainstream: British Press Coverage of Environmental Issues*, Chichester: Packard Publishing Ltd.

Watts, M. (1998) 'Nature as artifice and artifact', Chapter 11 in B. Braun and N. Castree (eds) *Remaking Reality: Nature at the Millennium*, London: Routledge.

Weizsacker, E. von, Lovins, A. and Lovins, L. (1998) *Factor Four: Doubling Wealth Halving Resource Use*, London: Earthscan.

Welford, R. (1994) *Environmental Strategy and Sustainable Development: The Corporate Challenge for the 21st Century*, London: Routledge.

Welford, R. (1997) *Hijacking Environmentalism: Corporate Responses to Sustainable Development*, London: Earthscan.

Welford, D. and Starkey, R. (eds) (1996) *Business and the Environment*, London: Earthscan.

Weller, J. (1967) *Modern Agriculture and Rural Planning*, London: Architectural Press.

Westmacott, R. and Worthington, T. (1976) *New Agricultural Landscapes*, Cheltenham: Countryside Commission.

Which? Magazine (1999) 'Surfing the High Street' January, 16–19.

Wibberley, G. (1982) *Countryside Planning: A Personal Evaluation*, Occasional Papers No. 7, Dept of Environmental Studies and Countryside Planning, Ashford: Wye College, University of London.

Wigan, M. (1991) *The Scottish Highland Estate: Preserving an Environment*, Shrewsbury: Swan Hill Press.

Williams, J.J. (1991) 'Meadowhall: its impact on Sheffield city centre and Rotherham', *International Journal of Retail and Distribution Management*, 19, 1, 29–37.

Williamson, T. and Bellamy, L. (1987) *Property and Landscape: A Social*

History of Land Ownership and the English Countryside, London: George Philip.
Wilson, G.A. (1994) 'German agri-environmental schemes: a preliminary review', *Journal of Rural Studies*, 10, 1, 27–46.
Wilson, J.F. (1995) *British Business History 1720–1994*, Manchester: Manchester University Press.
Winter, M. (1996) *Rural Politics: Policies for Agriculture, Forestry and the Environment*, London: Routledge.
Wood, C. (1995) *Environmental Impact Assessment: A Comparative View*, Harlow: Longman.
Woodell, S.R. (ed.) (1985) *The English Landscape: Past, Present and Future*, Oxford: Oxford University Press.
Woodland Trust (1997a) 'Regeneration at Howe Court', *Broadleaf*, 49, 1.
Woodland Trust (1997b) *Annual Review*, Grantham: The Woodland Trust.
Worcester, R. (1997) 'Public opinion and the environment', in M. Jacobs (ed.) *Greening the Millennium? The New Politics of the Environment*, Oxford: Blackwell.
World Resources Institute (1996) *World Resources: The Urban Environment 1996–7*, Oxford: Oxford University Press.
World Resources Institute (1998) *World Resources 1998–99: Environmental Change and Human Health*, Oxford: Oxford University Press.
Wrigley, N. (ed.) (1988) *Store Choice, Store Location and Market Analysis*, London: Routledge.
WWF (1999) 'NPI/WWF Investment Fund investing in companies of the future', *WWF News*, Spring, p. 3.

Index